Intel Edison Projects

Build exciting IoT projects with Intel Edison

Avirup Basu

BIRMINGHAM - MUMBAI

Intel Edison Projects

First published: May 2017

Production reference: 1250517

Published by Packt Publishing Ltd.
Livery Place
35 Livery Street
Birmingham
B3 2PB, UK.
ISBN 978-1-78728-840-9

www.packtpub.com

Credits

Author
Avirup Basu

Reviewer
Abhishek Nandy

Commissioning Editor
Vijin Boricha

Acquisition Editor
Prachi Bisht

Content Development Editor
Mamata Walkar

Technical Editor
Sayali Thanekar

Copy Editor
Safis Editing

Project Coordinator
Kinjal Bari

Proofreader
Safis Editing

Indexer
Aishwarya Gangawane

Graphics
Kirk D'Penha

Production Coordinator
Melwyn Dsa

About the Author

Avirup Basu is an independent developer based in Siliguri, West Bengal, India. His main areas of interest include IoT, robotics, and artificial intelligence. He holds a BTech degree in electronics and communication engineering from Siliguri Institute of Technology. He has been actively involved in projects involving robotics and IoT since his college days. He was a Microsoft student partner for four years from 2012-2016. In 2016, he was selected as an Intel Software Innovator. He has three research paper publications on computer vision and robotics, of which one is of IEEE on autonomous navigation and 2D mapping using SONAR. He holds seminars and workshops and has been training students and professionals in multiple areas, but mainly in IoT and robotics.

He has recently started uploading YouTube videos, where he focusses on topics related to IoT, robotics, application development, AI, and other fields of interest. The link for his channel is:

`https://www.youtube.com/user/Avirup171`

He can be contacted by e-mail at `avirup.basu@live.com` and his website is `http://www.avirupbasu.com/`.

I would like to express my gratitude to my beloved parents and my friends--without their constant support and motivation this would not have been possible. I would also like to thank my faculties at Siliguri Institute of Technology and colleagues at Altimetrik, whose motivation helped me a lot with this book.
Thanks to the Packt team, mainly Kinjal Bari, Prachi Bisht, Mamata Walker, and Sayali Thanekar, who have guided me through this book and helped me to be right on schedule, as well as the others attached to this book, and thanks to my technical reviewer, Abhishek Nandy, without whom this would not have been possible.

About the Reviewer

Abhishek Nandy is a software developer, innovator, and community speaker. He has experience in AI, IoT, game development, desktop development, web development, cloud, and Android. Abhishek is also the winner of popular choice Hack Productivity for Hololens Office 365 bot, and was a top 50 Innovators for Digital India Innovate. He was trained at IIMA, an application architect at Prescriber 360, founder of Geek Monkey Studios, AI trainer, and consultant. He has provided training on Intel AI at prestigious colleges such as IITR, IITG, DIT, and UPES, Dehradun. He has been awarded Microsoft MVP Development Platform, Intel Black Belt Developer, and Intel Software Innovator.

Abhishek has worked on books such as *Beginning Platino Engine and Leap Motion for Developers* by Apress.

> *I would like to thank Anupam Nandwana and Edward Vaz, who gave me the opportunity to perform a full-fledged experiment for a project on the pharama industry using AI and IoT, and currently piloting the project from scratch. I would like to thank Sourav Lahoti, who also worked with me in finalizing the product, who is a good friend of mine and an excellent human being and, more than that, a geek who has a lot of knowledge to share and is a very good learner.*
> *I would like to thank my parents, without whom I couldn't have achieved anything.*

www.PacktPub.com

For support files and downloads related to your book, please visit www.PacktPub.com.

Did you know that Packt offers eBook versions of every book published, with PDF and ePub files available? You can upgrade to the eBook version at www.PacktPub.comand as a print book customer, you are entitled to a discount on the eBook copy. Get in touch with us at service@packtpub.com for more details.

At www.PacktPub.com, you can also read a collection of free technical articles, sign up for a range of free newsletters and receive exclusive discounts and offers on Packt books and eBooks.

https://www.packtpub.com/mapt

Get the most in-demand software skills with Mapt. Mapt gives you full access to all Packt books and video courses, as well as industry-leading tools to help you plan your personal development and advance your career.

Why subscribe?

- Fully searchable across every book published by Packt
- Copy and paste, print, and bookmark content
- On demand and accessible via a web browser

Customer Feedback

Thanks for purchasing this Packt book. At Packt, quality is at the heart of our editorial process. To help us improve, please leave us an honest review on this book's Amazon page at `https://www.amazon.com/dp/1787288404`.

If you'd like to join our team of regular reviewers, you can e-mail us at `customerreviews@packtpub.com`. We award our regular reviewers with free eBooks and videos in exchange for their valuable feedback. Help us be relentless in improving our products!

Table of Contents

Preface

Intel Edison Projects is meant for beginners who want to get to grips with the Intel Edison and explore its features. Intel Edison is an embedded computing platform, which allows us to explore areas of IoT, embedded systems, and robotics.

This book takes you through various concepts and each chapter has a project that can be performed by you. It covers multiple topics, including sensor data acquisition and pushing it to the cloud to control devices over the Internet, as well as topics ranging from image processing to both autonomous and manual robotics.

In every chapter, the book first covers some theoretical aspects of the topic, which include some small chunks of code and a minimal hardware setup. The rest of the chapter is dedicated to the practical aspects of the project.

The projects discussed in this book wherever possible require only minimal hardware, and the projects in each chapter are included to make sure that you understand the basics.

What this book covers

Chapter 1, *Setting up Intel Edison*, covers the initial steps of setting up the Intel Edison, flashing it, and setting up the environment for development.

Chapter 2, *Weather Station (IoT)*, introduces you to IoT and uses a simple case of a weather station where we use temperature, smoke level, and sound level and push data to the cloud to visualize it.

Chapter 3, *Intel Edison and IoT (Home Automation)*, covers a case for home automation, where we are controlling electrical load using the Intel Edison.

Chapter 4, *Intel Edison and Security System*, covers voice and image processing for the Intel Edison.

Chapter 5, *Autonomous Robotics with Intel Edison*, explores the field of robotics, where we develop a line-following robot using the Intel Edison and related algorithms.

Chapter 6, *Manual Robotics with Intel Edison*, explores UGVs and also guides you through the process of developing controller software.

What you need for this book

The mandatory prerequisites for this book are the Intel Edison with Windows/Linux/Mac OS. The software requirements are as follows:

- Arduino IDE
- Visual Studio
- FileZilla
- Notepad++
- PuTTY
- Intel XDK

Who this book is for

If you are a hobbyist, robot engineer, IoT enthusiast, programmer, or developer who wants to create autonomous projects with the Intel Edison, then this book is for you. Prior programming knowledge would be beneficial.

Conventions

In this book, you will find a number of text styles that distinguish between different kinds of information. Here are some examples of these styles and an explanation of their meaning.

Code words in text, database table names, folder names, filenames, file extensions, pathnames, dummy URLs, user input, and Twitter handles are shown as follows: "We can include other contexts through the use of the `include` directive."

A block of code is set as follows:

```
int a = analogRead(tempPin ); float R = 1023.0/((float)a)-1.0;
R = 100000.0*R;

float temperature=1.0/(log(R/100000.0)/B+1/298.15)-273.15;
Serial.print("temperature = "); Serial.println(temperature);
delay(500);
```

When we wish to draw your attention to a particular part of a code block, the relevant lines or items are set in bold:

```
string res = textBox.Text; if(string.IsNullOrEmpty(res))
  {
    MessageBox.Show("No text entered. Please enter again");
```

```
    }
else
  {
    textBlock.Text = res;
```

Any command-line input or output is written as follows:

```
npm install mqtt
```

New terms and **important words** are shown in bold. Words that you see on the screen, for example, in menus or dialog boxes, appear in the text like this: "After you click on **OK**, the tool will automatically unzip the file."

Warnings or important notes appear in a box like this.

Tips and tricks appear like this.

Reader feedback

Feedback from our readers is always welcome. Let us know what you think about this book-what you liked or disliked. Reader feedback is important for us as it helps us develop titles that you will really get the most out of.

To send us general feedback, simply e-mail feedback@packtpub.com, and mention the book's title in the subject of your message.

If there is a topic that you have expertise in and you are interested in either writing or contributing to a book, see our author guide at www.packtpub.com/authors.

Customer support

Now that you are the proud owner of a Packt book, we have a number of things to help you to get the most from your purchase.

Downloading the example code

You can download the example code files for this book from your account at `http://www.p acktpub.com`. If you purchased this book elsewhere, you can visit `http://www.packtpub.c om/support` and register to have the files e-mailed directly to you.

You can download the code files by following these steps:

1. Log in or register to our website using your e-mail address and password.
2. Hover the mouse pointer on the **SUPPORT** tab at the top.
3. Click on **Code Downloads & Errata**.
4. Enter the name of the book in the **Search** box.
5. Select the book for which you're looking to download the code files.
6. Choose from the drop-down menu where you purchased this book from.
7. Click on **Code Download**.

Once the file is downloaded, please make sure that you unzip or extract the folder using the latest version of:

- WinRAR / 7-Zip for Windows
- Zipeg / iZip / UnRarX for Mac
- 7-Zip / PeaZip for Linux

The code bundle for the book is also hosted on GitHub at `https://github.com/PacktPubl ishing/Intel-Edison-Projects`. We also have other code bundles from our rich catalog of books and videos available at `https://github.com/PacktPublishing/`. Check them out!

Downloading the color images of this book

We also provide you with a PDF file that has color images of the screenshots/diagrams used in this book. The color images will help you better understand the changes in the output. You can download this file from `https://www.packtpub.com/sites/default/files/down loads/IntelEdisonProjects_ColorImages.pdf`.

Errata

Although we have taken every care to ensure the accuracy of our content, mistakes do happen. If you find a mistake in one of our books-maybe a mistake in the text or the code-we would be grateful if you could report this to us. By doing so, you can save other readers from frustration and help us improve subsequent versions of this book. If you find any errata, please report them by visiting http://www.packtpub.ccm/submit-errata, selecting your book, clicking on the **Errata Submission Form** link, and entering the details of your errata. Once your errata are verified, your submission will be accepted and the errata will be uploaded to our website or added to any list of existing errata under the Errata section of that title.

To view the previously submitted errata, go to https://www.packtpub.com/books/content/support and enter the name of the book in the search field. The required information will appear under the **Errata** section.

Piracy

Piracy of copyrighted material on the Internet is an ongoing problem across all media. At Packt, we take the protection of our copyright and licenses very seriously. If you come across any illegal copies of our works in any form on the Internet, please provide us with the location address or website name immediately so that we can pursue a remedy.

Please contact us at copyright@packtpub.com with a link to the suspected pirated material.

We appreciate your help in protecting our authors and our ability to bring you valuable content.

Questions

If you have a problem with any aspect of this book, you can contact us at questions@packtpub.com, and we will do our best to address the problem.

1
Setting up Intel Edison

In every **Internet of Things (IoT)** or robotics project, we have a controller that is the brain of the entire system. Similarly, we have the Intel Edison. The Intel Edison computing module comes in two different packages: one is a mini-breakout board; the other is an Arduino-compatible board. One can use the board in its native state as well, but in that case the we have to fabricate our own expansion board. The Edison is basically the size of an SD card. Due to its tiny size, it's perfect for wearable devices. However, it's capabilities makes it suitable for IoT applications; and above all, the powerful processing capability, makes it suitable for robotics applications. However we don't simply use the device in this state. We hook up the board with an expansion board. The expansion board provides the user with enough flexibility and compatibility for interfacing with other units. The Edison has an operating system that runs the entire system. It runs a Linux image. So, to set up your device, you initially need to configure your device both at the hardware and at the software level.

In this chapter, we will be covering the following topics:

- Setting up the Intel Edison
- Setting up the developer environment
- Running sample programs on the board using Arduino IDE, Intel XDK, and others
- Interacting with the board by using our PC

Initial hardware setup

We'll concentrate on the Edison package that comes with an Arduino expansion board.
Initially, you will get two different pieces:

- The Intel® Edison board
- The Arduino expansion board

The following figure shows the architecture of the device:

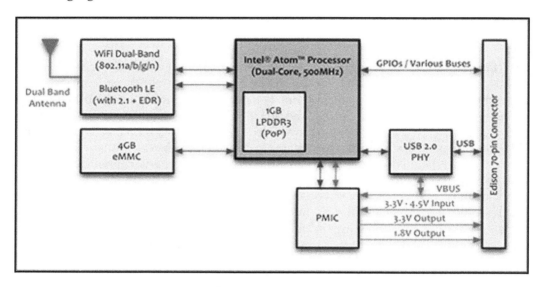

Architecture of Intel Edison. Picture Credits: http://www.software.intel.com

We need to hook these two pieces up in a single unit. Place the Edison board on top of the
expansion board so that the GPIO interfaces meet at a single point. Gently push the Edison
against the expansion board. You will hear a click. Use the screws that come with the
package to tighten the setup. Once this is done, we'll now set up the device both at
hardware level and software level to be used further. The following are the steps we'll cover
in detail:

1. Downloading the necessary software packages
2. Connecting your Intel® Edison to your PC
3. Flashing your device with the Linux image
4. Connecting to a Wi-Fi network
5. SSH-ing your Intel® Edison device

Downloading the necessary software packages

To move forward with the development on this platform, we need to download and install a couple of software packages, which includes the drivers and the IDEs. The following is the list of the software along with the links that are required:

- Intel® Platform Flash Tool Lite (`https://01.org/android-ia/downloads/intel-platform-flash-tool-lite`)
- PuTTY (`http://www.chiark.greenend.org.uk/~sgtatham/putty/download.html`)
- Intel XDK for IoT (`https://software.intel.com/en-us/intel-xdk`)
- Arduino IDE (`https://www.arduino.cc/en/Main/Software`)
- FileZilla FTP client (`https://filezilla-project.org/download.php`)
- Notepad ++ or any other editor (`https://notepad-plus-plus.org/download/v7.3.html`)

Drivers and miscellaneous downloads

Drivers and miscellaneous can be downloaded from:

- Latest Yocto Poky image
- Windows standalone driver for the Intel Edison
- FTDI drivers (`http://www.ftdichip.com/Drivers/VCP.htm`)

The first and the second packages can be downloaded from `https://software.intel.com/en-us/iot/hardware/edison/downloads`.

Plugging in your device

After the software and drivers have all been installed, we'll connect the device to a PC. You need two Micro-B USB cables(s) to connect your device to the PC. You can also use a 9V power adapter and a single Micro-B USB cable, but for now we won't use the power adapter. The main use of the power adapter will come in a later section of this book, especially when we'll be interfacing with other devices that require USB.

The following figure shows different sections of an Arduino expansion board of the Intel Edison:

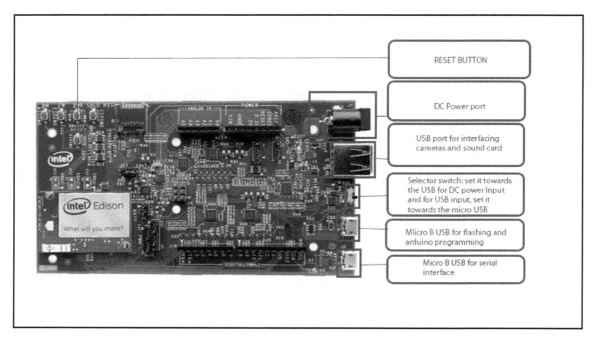

Different sections of an Arduino expansion board of Intel Edison

A small switch exists between the USB port and the OTG port. This switch must be towards the OTG port because we're going to power the device from the OTG port and not through the DC power port. Once it is connected to your PC, open your device manager and expand the ports section. If all the installations of the drivers were successful, then you'll see two ports:

- Intel Edison virtual com port
- USB serial port

Flashing your device

Once your device is successfully detected and installed, you need to flash your device with the Linux image. For this we'll use the flash tool provided by Intel:

1. Open the flash lite tool and connect your device to the PC:

Intel phone flash lite tool

2. Once the flash tool is opened, click on **Browse...** and browse to the .zip file of the Linux image you have downloaded.
3. After you click on **OK**, the tool will automatically unzip the file.

4. Next, click on **Start** to flash:

Intel® Phone flash lite tool — stage 1

5. You will be asked to disconnect and reconnect your device. Do this, and the board should start flashing. It may take some time before the flashing is completed. Don't tamper with the device during this process.

6. Once the flashing is completed, we can configure the device:

Intel® Phone flash lite tool — complete

Configuring the device

After flashing successfully, we'll now configure the device. We're going to use the PuTTY console for the configuration. PuTTY is an SSH and telnet client, developed originally by Simon Tatham for the Windows platform. We're going to use the **Serial** section here.

Before opening the PuTTY console, open up the **Device manager** and note the port number for the USB serial port. This will be used in your PuTTY console:

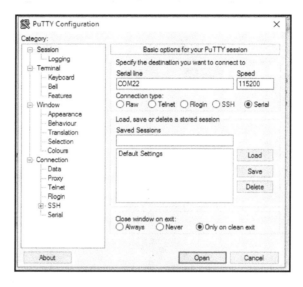

Ports for Intel® Edison in PuTTY

Next, select **Serial** on the PuTTY console and enter the port number. Use a baud rate of 115,200. Press **Open** to open the window for communicating with the device:

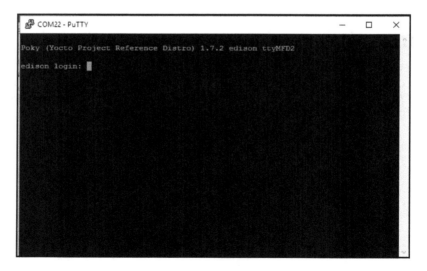

PuTTY console — login screen

Once you are in the PuTTY console, you can execute commands to configure your Edison. The following is the set of tasks we'll do in the console to configure the device:

1. Provide a name for your device.
2. Provide a root password (SSH your device).
3. Connect your device to Wi-Fi.

Initially, when in the console, you will be asked to log in. Type in root and press *Enter*. You will see **root@edison**, which means that you are in the root directory:

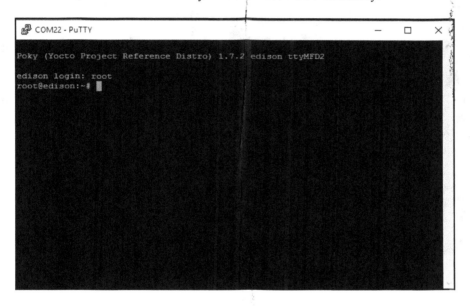

PuTTY console — login success

Now, we are in the Linux Terminal of the device. Firstly, we'll enter the following command for the setup:

```
configure_edison –setup
```

Press *Enter* after entering the command, and the entire configuration will be straightforward:

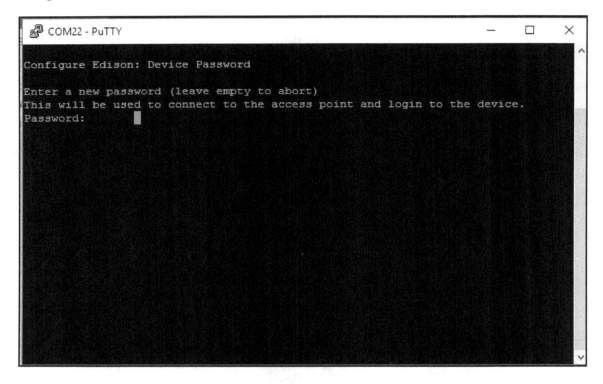

PuTTY console — set password

Firstly, you will be asked to set a password. Type in a password and press *Enter*. You need to type in your password again for confirmation. Next, we'll set up a name for the device:

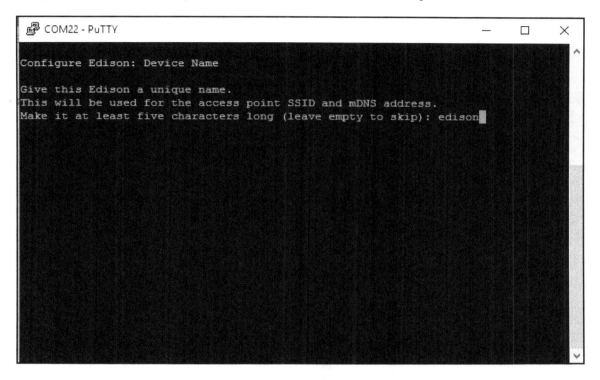

PuTTY console — set name

Give a name for your device. Please note that this is not the login name for your device. It's just an alias for your device. Also the name should be atleast five characters long. Once you've entered the name, it will ask for confirmation: press *y* to confirm. Then it will ask you to set up Wi-Fi. Again select *y* to continue. It's not mandatory to set up Wi-Fi, but it's recommended. We need the Wi-Fi for file transfer, downloading packages, and so on:

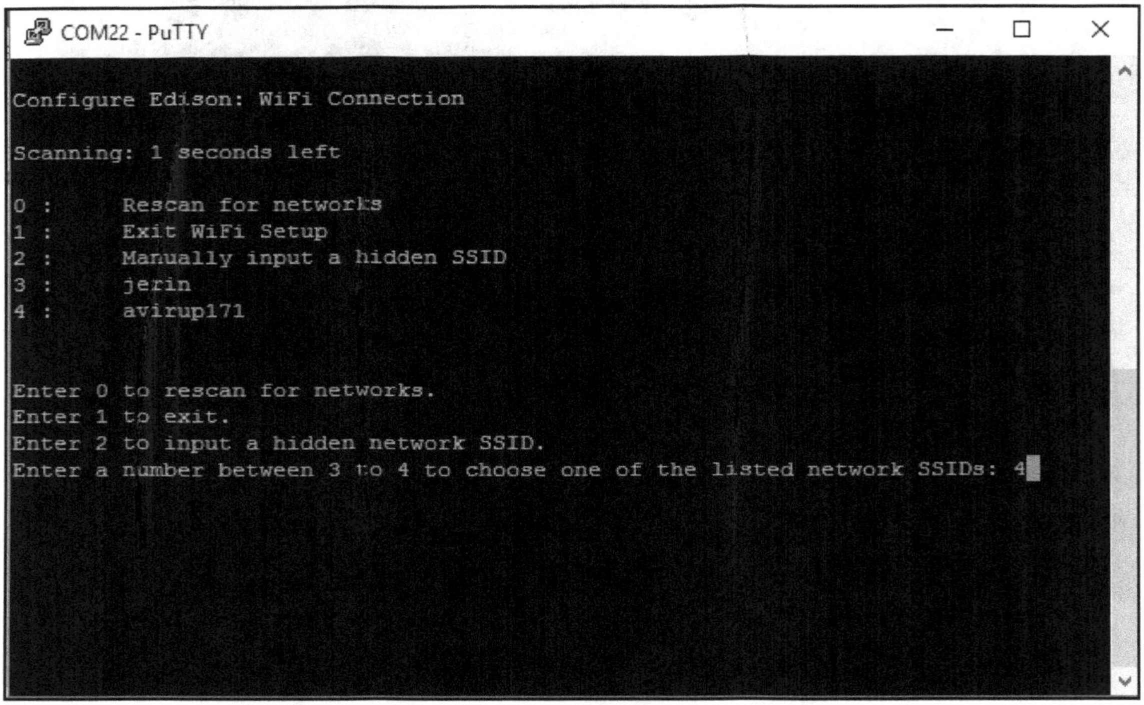

```
COM22 - PuTTY                                           —    □    ×

Configure Edison: WiFi Connection

Scanning: 1 seconds left

0 :       Rescan for networks
1 :       Exit WiFi Setup
2 :       Manually input a hidden SSID
3 :       jerin
4 :       avirup171


Enter 0 to rescan for networks.
Enter 1 to exit.
Enter 2 to input a hidden network SSID.
Enter a number between 3 to 4 to choose one of the listed network SSIDs: 4
```

PuTTY console — set Wi-Fi

Once the scanning is completed, we'll get a list of available networks. Select the number corresponding to your network and press *Enter*. In this case, it's 5, which corresponds to **avirup171** which is my Wi-Fi. Enter the network credentials. After you do that, your device will be connected to Wi-Fi. You should get an IP address after your device is connected:

```
COM22 - PuTTY                                        —    □    ×

Scanning: 1 seconds left

0 :      Rescan for networks
1 :      Exit WiFi Setup
2 :      Manually input a hidden SSID
3 :      blrAirtel
4 :      Airtelbina
5 :      avirup171


Enter 0 to rescan for networks.
Enter 1 to exit.
Enter 2 to input a hidden network SSID.
Enter a number between 3 to 5 to choose one of the listed network SSIDs: 5
Is avirup171 correct? [Y or N]: y
Password must be between 8 and 63 characters.
What is the network password?: **********
Initiating connection to avirup171. Please wait...
Attempting to enable network access, please check 'wpa_cli status' after a minut
e to confirm.
Done. Please connect your laptop or PC to the same network as this device and go
 to http://192.168.0.101 or http://edison.local in your browser.
root@edison:~#
```

PuTTY console — set Wi-Fi -2

After successful connection, you should get this screen. Make sure your PC is connected to the same network. Open up the browser in your PC, and enter the IP address shown in the console. You should get a screen similar to this:

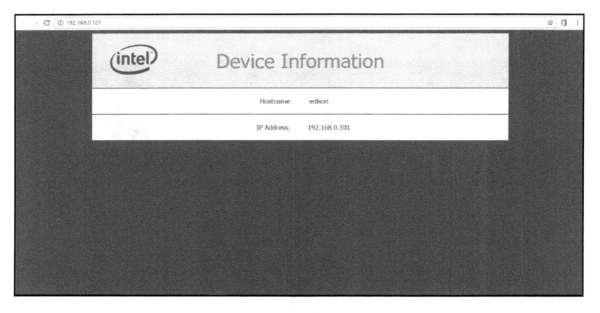

Wi-Fi setup — completed

Now, we've finished with the initial setup. However, the Wi-Fi setup normally doesn't happen in one go. Sometimes your device doesn't get connected to Wi-Fi and sometimes we cannot get the page shown previously. In those cases, you need to start `wpa_cli` to manually configure Wi-Fi.

Refer to the following link for the details:

```
http://www.intel.com/content/www/us/en/support/boards-and-kits/000006202.html
```

With Wi-Fi setup completed, we can move forward to set up our developer environment. We'll cover the following programming languages and the respective IDEs:

- Arduino processor language (C/C++)
- Python
- Node.js

Arduino IDE

The Arduino IDE is a famous, and widely used, integrated developer environment that not only covers Arduino boards but also many other boards of Intel including Galileo, Edison, Node MCU, and so on. The language is based on C/C++. Once you download the Arduino IDE from the link mentioned at the beginning of this chapter, you may not receive the Edison board package. We need to manually download the package from the IDE itself. To do that, open up your Arduino IDE, and then go to **Tools** | **Board: "Arduino/Genuino Uno"** | **Board Manager...**:

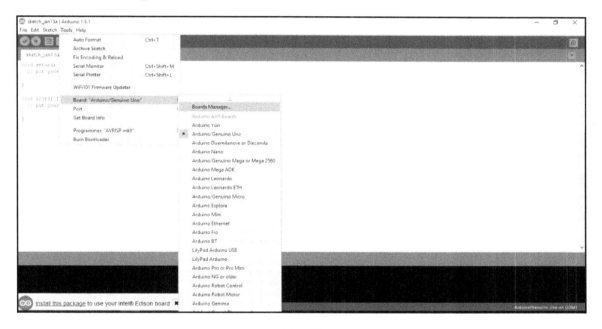

Arduino IDE

You now need to click on **Boards Manager** and select **Intel i686 Boards**. Click on the version number and then click on **Install**. **Boards Manager** is an extremely important component of the IDE. We use the **Boards Manager** to add external Arduino-compatible boards:

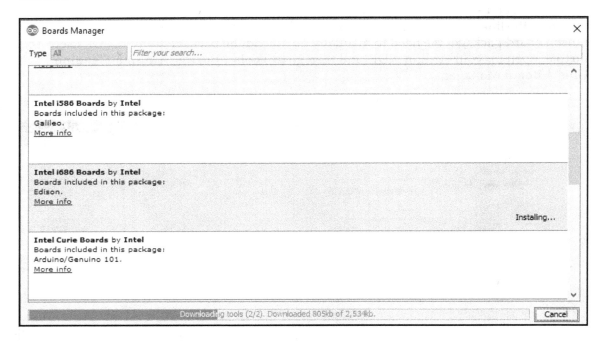

Boards Manager

Once installed, you should see your board displayed under **Tools Boards**:

Board installation successful

Once successfully installed, you will now be able to program the device using the IDE. Like every starter program, we'll also be burning a simple program into the Intel Edison which will blink the on-board LED at certain intervals set by us. Through this, the basic structure of the program using the Arduino IDE will also be clear. When we initially open the IDE, we get two functions:

- `void setup()`
- `void loop()`

The setup function is the place where we declare whether the pins are to be configured in output mode or input mode. We also start various other services, such as serial port communication, in the setup method. Depending on the usecase, the implementation changes. The loop method is that segment of the code that executes repeatedly in an infinite sequence. Our main logic goes in here. Now we need to blink an LED with an interval of 1 second:

```
#define LED_PIN 13
void setup()
  {
    pinMode(LED_PIN, OUTPUT);
  }

void loop()
  {
    digitalWrite(LED_PIN, HIGH);
    delay(1000);
    digitalWrite(LED_PIN, LOW);
    delay(1000);
  }
```

In the preceding code, the line `#define LED_PIN 13` is a macro for defining the LED pin. In the Arduino expansion board, an LED and a resistor is already attached to `pin 13`, so we do not need to attach any additional LEDs for now. In the setup function, we have defined the configuration of the pin as output using the pinMode function with two parameters. In the loop function, we have initially set the pin to high by using the `digitalWrite` function with two parameters, and then we've defined a delay of 1,000 miliseconds which is equivalent of 1 second. After the delay, we set the pin to low and then again define a delay of 1 second. The preceding code explains the basic structure of the Arduino code written in the Arduino IDE.

To burn this program to the Edison device, first compile the code using the **compile** button, then select the port number of your device, and finally click the **Upload** button to upload the code:

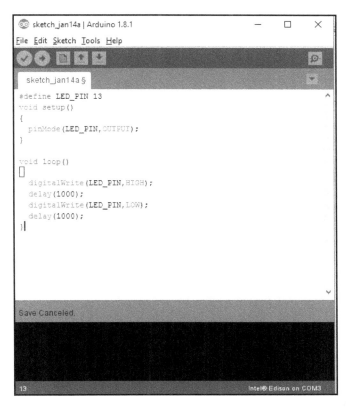

Arduino IDE — blink

The port number can be selected under **Tools** | **port**.

Now that we know how to program using an Arduino, let's have a look at how it actually works or what's happening inside the Arduino IDE.

A number of steps actually happen while uploading the code:

1. First, the Arduino environment performs some small transformations to make sure that the code is correct C or C++ (two common programming languages).
2. It then gets passed to a compiler (`avr-gcc`), which turns the human readable code into machine readable instructions (or object files).

3. Then, your code gets combined with (linked against), the standard Arduino libraries that provide basic functions such as `digitalWrite()` or `Serial.print()`. The result is a single Intel hex file, which contains the specific bytes that need to be written to the program memory of the chip on the Arduino board.

4. This file is then uploaded to the board, transmitted over the USB or serial connection via the bootloader already on the chip or with external programming hardware.

Python

Edison can also be programmed in Python. The code needs to be run on the device directly. We can either directly program the device, using any editor, such as the VI editor, or write the code in the PC first, and then transfer it using any FTP client, like FileZilla. Here we'll first write the code using Notepad++ and then transfer the script. Here also, we'll be executing a simple script which will blink the on-board LED. While dealing with Python and hardware, we need to use the MRAA library to interface with the GPIO pins. This is a low-level skeleton library for communication on GNU/Linux platforms. It supports almost all of the widely-used Linux-based boards. So, initially you need to install the library on the board.

Open up PuTTY and log in to your device. Once logged in, we'll add AlexT's unofficial `opkg` repository.

To do that, add the following lines to `/etc/opkg/base-feeds.conf` using the VI editor:

```
src/gz all http://repo.opkg.net/edison/repo/all
src/gz edison http://repo.opkg.net/edison/repo/edison
src/gz core2-32 http://repo.opkg.net/edison/repo/core2-32
```

Next, `update` the package manager and `install git` by executing the following commands:

```
opkg update
opkg install git
```

We'll clone Edison-scripts from GitHub to simplify certain things:

```
git clone https://github.com/drejkim/edison-scripts.git ~/edison-scripts
```

Next we'll add ~/edison-scripts to the path:

```
echo 'export PATH=$PATH:~/edison-scripts' >> ~/.profile
source ~/.profile
```

We'll now run the following scripts to complete the process. Please note that the previous steps will not only configure the device for MRAA, but will also set up the environment for later projects in this book.

Firstly, run the following script. Just type:

```
resizeBoot.sh
Then go for
installPip.sh
```

The previous package is the Python package manager. This will be used to install essential Python packages to be used in a later part of this book. Finally, we'll install **Mraa** by executing the following command:

```
installMraa.sh
```

MRAA is a low-level skeleton library for communication on GNU/Linux platforms. Libmraa is a C/C++ library with bindings to Java, Python, and JavaScript to interface with the IO on Galileo, Edison, and other platforms. In simple words, it allows us to operate on the IO pins.

Once the preceding steps have completed, we are good to go with the code for Python. For that, open up any code editor, such as Notepad++, and type in the following code:

```
import mraa
import time


led = mraa.Gpio(13)
led.dir(mraa.DIR_OUT)

while True:
        led.write(1)
        time.sleep(0.2)
        led.write(0)
        time.sleep(0.2)
```

Please save the preceding code as a .py extension such as blink.py, and now, we'll explain it line by line.

Initially, using the import statements, we import two libraries: MRAA and time. MRAA is required for interfacing with the GPIO pins:

```
led = mraa.Gpio(13)
led.dir(mraa.DIR_OUT)
```

Here we initialize the LED pin and set it to the output mode:

```
while True:
        led.write(1)
        time.sleep(0.2)
        led.write(0)
        time.sleep(0.2)
```

In the preceding block, we put our main logic in an infinite loop block. Now, we will transfer this to our device. To do that again, go to the PuTTY console and type `ifconfig`. Under the `wlan0` section, note down your IP address:

IP address to be used

Now open up FileZilla and enter your credentials. Make sure your device and your PC are on the same network:

- **Host**: The IP address you got according to the preceding screenshot: `192.168.0.101`
- **Username**: `root` because you will be logging in to the root directory
- **Password**: Your Edison password
- **Port**: `22`

Once entered, you will get the folder structure of the device. We'll now transfer the Python code from our PC to the device. To do that, just locate your `.py` file in Windows Explorer and drag and drop the file in the FileZilla console's Edison's folder. For now, just paste the file under `root`. Once you do that and if it's a success, the file should be visible in your Edison device by accessing the PuTTY console and executing the `ls` command.

Another alternative is to locate your file on the left-hand side of FileZilla; once located, just right-click on the file and click **Upload**. The following is the typical screenshot of the FileZilla windows:

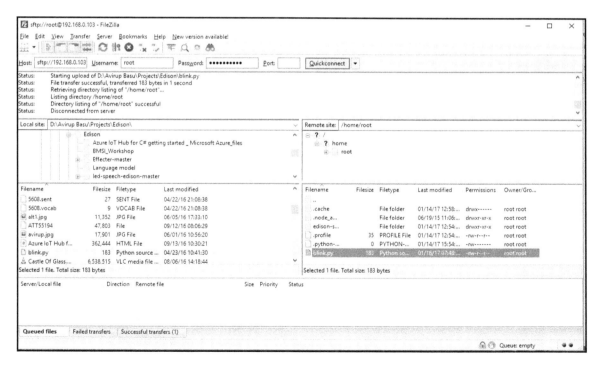

FileZilla application

Once transferred and successfully listed using the `ls` command, we are going to run the script. To run the script, in the PuTTY console, go to your `root` directory and type in the following command:

```
python blink.py
```

If the file is present, then you should get the LED blinking on your device. Congrats! You have successfully written a Python script on your Edison board.

Intel XDK for IoT (Node.js)

Another IDE we will be covering is the powerful cross-platform development tool by Intel: Intel XDK. This will be used to run our Node.js scripts. Ideally we run our Node.js scripts from the XDK, but there is always an option to do the same by just transferring the `.js` file to your device using an FTP client such as FileZilla and use node `FileName.js` to run your script. From the list of downloaded software provided at the beginning of this chapter, download and install the XDK and open it. You may be required to sign in to the Intel developer zone. Once done, open your XDK. Then, under IoT embedded applications, select a **Blank IoT Node.js Template**:

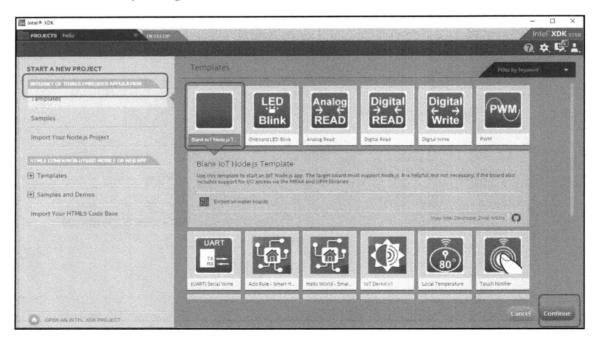

Screenshot for XDK

Once opened, replace all the existing code with the following code:

```
var m = require('mraa'); //require mraa
console.log('MRAA Version: ' + m.getVersion()); //write the mraa version to
the console

varmyLed = new m.Gpio(13); //LED hooked up to digital pin 13 (or built in
pin on Galileo Gen1 & Gen2 or Edison)
myLed.dir(m.DIR_OUT); //set the gpio direction to output
varledState = true; //Boolean to hold the state of Led

functionperiodicActivity()
  {
    myLed.write(ledState?1:0);
    ledState = !ledState;
    setTimeout(periodicActivity,1000);
  }
periodicActivity(); //call the periodicActivity function
```

If you have a close look at the code, then you may notice that the structure of the code remains more or less similar as that of the other two platforms. We initially import the MRAA library:

```
var m = require('mraa');
console.log('MRAA Version: ' + m.getVersion());
```

We also display the version of MRAA installed (you can skip this step). The next task is to initialize and configure the pin to be in output or input mode:

```
varmyLed = new m.Gpio(13);
myLed.dir(m.DIR_OUT);
varledState = true;
```

We use `ledState` to get the present state of the LED. Next, we define the logic in a separate function for blinking:

```
functionperiodicActivity()
  {
    myLed.write(ledState?1:0);
    ledState = !ledState;
    setTimeout(periodicActivity,1000);
  }
periodicActivity();
```

Finally, we call the function. On close inspection of the code, it's evident that the we have used only one delay in milliseconds as we are checking the present state using the tertiary operator. In order to execute the code on the device, we need to connect our device first.

To connect your device to the XDK, go to the **IoT Device** section, and click on the dropdown. You may see your device in the dropdown. If you see it, then click on **Connect**:

XDK screenshot — connection pane

If the device is not listed, then we need to add a manual connection. Click on **Add Manual Connection**, then add the credentials:

Screenshot for manual connection

In the address, put in the IP which was used in FileZilla. In the **Username**, insert `root`, and the password is the password that was set before. Click on **Connect** and your device should be connected. Click on **Upload** to upload the program and **Run** to run the program:

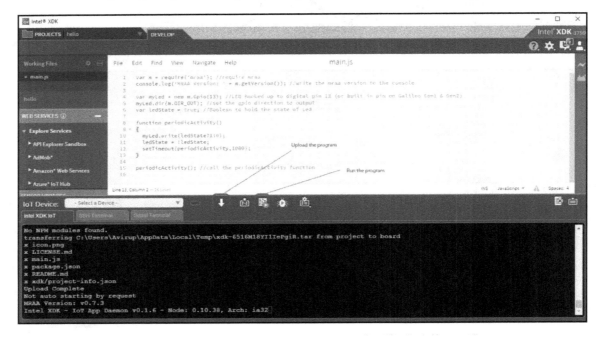

Screenshot for uploading and executing the code

After uploading, the LED that is attached to pin `13` should blink. Normally, when dealing with complex projects, we go for blank templates so that it's easy to customize and do the stuff we need.

For more examples and details on the XDK are available at: `https://software.intel.com/en-us/getting-started-with-xdk-and -iot`

Summary

In this chapter, we've covered the initial setup of the Intel Edison and configuring it to the network. We have also looked at how to transfer files to and from the Edison, and set up the developer environment for Arduino, Python, and Node.js. We did some sample programming, blinking an LED, using all three platforms. Through this, we've gained a fair knowledge of operating the Edison and developing simple to complex projects.

In Chapter 2, *Weather Station (IoT)*, we'll build a mini-weather station and will be able to deploy a project on the Intel Edison.

2
Weather Station (IoT)

In `Chapter 1`, *Setting up Intel Edison*, we learned how to set up the Intel Edison and run some basic programs in various languages. In this chapter, we'll scale up things a bit and introduce the Internet to the picture of the IoTs. We will hear this term quite often, and there has been no exact definition of it. However, if we try to bring out the literal meaning of IoT, it means connecting devices to the Internet. It may also be referred to as **connected devices** or **smart devices**. It's an advanced form of **machine to machine (M2M)** communication. In this chapter, the main focus will be on the architecture of IoTs and how Intel Edison is the perfect choice for developing systems revolving around IoT. We'll also be dealing with various components of a typical IoT project and how things can be scaled up at the component level. By the end of this chapter, we'll have built a system using Intel Edison that will take in pollution data, temperature data, and sound data, and upload it to the Web and display it live on a dashboard.

In this chapter, we will cover the following topics:

- Architecture of a typical IoT system interfacing sensors with Intel Edison
- Connecting the device and uploading data to the cloud deploying a mini weather station with Intel Edison

Overview of IoT and its usage

IoT has lots of usage and can make normal things behave smartly. Starting from industries, where we deal with large machinery, such as industrial robots and assembly line industrial units, IoT has its roots embedded in many uses providing vital data to the cloud, or a local server, where we can monitor them remotely and provide remote functionality in every place that uses IoT. Now imagine a healthcare solution where we have an aged member of our family and we need to monitor them regularly.

Smart medical devices come to the picture, where the data collected from the human body is constantly pushed to the cloud and monitored at a local level for any signs of abnormality. Proper analytics is performed at cloud level where using machine learning algorithms, we can predict the health of the person. Considering the use case of home automation that will be covered in `Chapter 3`, *Intel Edison and IoT (Home Automation)*, we can deal with controlling devices and monitoring the usage of devices such as electrical load such as lights, fans, and AC, remotely from anywhere in the world.

Architecture of a typical IoT project

When dealing with IoT projects, there are certain key factors that should be kept in mind, some of which are as follows:

- Choice of hardware
- Choice of the networking protocol
- Choice of the sensors
- Choice of the IoT platform

Beside the previously mentioned points, several other factors come into play that will be clear later in the chapter. When dealing with IoT, the first thing that comes to our mind is uploading data. Typically, what is involved in an IoT project is we take readings from sensors and upload them to the Web. In this process, several sub-processes, such as processing and filtering data, analytics, and machine learning, come to play.

The following diagram depicts a standard architecture:

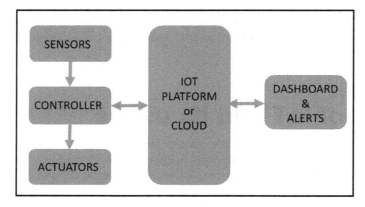

IoT architecture overview

As depicted in the preceding diagram, the typical IoT system may consist of these components:

- Sensors
- Controller actuators
- IoT platform or cloud
- Dashboard and alerts

Sensors

Sensors form a key role in the IoT space. These act as a kind of connection layer between the physical world and the digital world. The sensors coupled provide the entire system with the data that we want. Currently we aren't focused on industrial-grade sensors, but low-power sensors are suitable for small proof of concepts. When dealing with sensors, certain factors should be kept in mind:

- Power used by the sensor
- Type of output provided (digital or analog)
- Sensitivity of the sensor

The sensor usually performs one-way communication with the controller. The data captured by the sensors is converted into digital values using an ADC, and we obtain the output through a controller. Let's consider a use case of a light sensor. In this case, the sensor returns a calibrated value of the resistance. Thus we get a resistance value based on the light intensity. It is similar for other sensors, where any physical component is read and converted to a digital component. Normally a sensor has three pins; Vcc, Gnd, and signal. The interfacing of sensors with the controller will be discussed later in the chapter.

Controllers

These are the brain of an IoT system. Controllers perform most of the action in the IoT space. In industry, controllers may be referred as IoT gateways. From data acquisition to processing this data to uploading it to the cloud and control actuators, are all performed by the controller. Usually, controllers have the capabilities of networking and wireless communication. If some controllers don't have that capability, then usually they are stacked up with external networking devices. Arduino, Raspberry Pi, Intel Galileo, and Intel Edison are some of the most commonly used controllers. Controllers normally come with GPIO pins that interface with other devices. Most of the details of Intel Edison, the controller that we are using, were discussed in `Chapter 1`, *Setting up Intel Edison*.

Actuators

Actuators consist of mainly electro-mechanical devices that provide us some action on the basis of controller signals. Based on signals sent by the controllers, actuators get activated or deactivated. Actuators are mainly motors; servos fall into this category as well. Controllers cannot directly control actuators because of the high power requirements or the nature of the current (AC or DC) of the actuators. The interfacing hardware component is called a driver circuit.

Controllers send control signals to the driver circuits and based on these control signals, the actuators gets activated. While dealing with the robotics module of this book, the actuators will be discussed in detail.

Cloud or IoT platform

This is perhaps one of the most important parts of the entire IoT ecosystem because of the fact that at the end we need to upload the data so that it can be accessed from anywhere, which is the Internet part of IoTs. Normally we prefer to go for IoTs platforms. These platforms have their own SDKs and we just use those SDKs to pair up our device with the IoT platform and upload the data. Top-level players in the IoT platform include Microsoft Windows Azure, IBM Bluemix, and Amazon Web Services. There are other platforms that provide those services, such as Datonis, Dweet, Thingspeak, Thingworx, and many more. The choice of the IoT platform is very specific to the use case depending on certain factors, such as:

- **Protocols it supports**: These are mainly REST, web sockets, and MQTT
- **Dashboard capability**: The platforms, own dashboards for visualization and the flexibility of development of custom dashboards
- **Rule engines**: These include the rules that are needed to be defined based on the incoming data
- **Event-based services**: These are the necessity of triggering of events based on the output of the rule engines

In this chapter, we will be discussing the use of `dweet.io` for our project, a mini weather station.

Dashboards and alerts

When we have data, the data should be viewed in the form of gauges, graphs, and so on. In most cases, the platform itself provides support, but sometimes there arises a requirement of building custom dashboard based on the data we get. Normally through REST API calls, we get the data from the IoT platform to our custom dashboards. These are also essential components of an IoT ecosystem. Dashboards must also be compatible with devices ranging from PCs to mobile devices. They also must be in real time. While dashboards deal with data visualizations, alerts on the other hand are responsible for notifying the user of any malfunctions or any abnormalities in the system. Normally the preferred mode of alerts is either some visualizations, push notifications, e-mails, text messages, and so on.

Interfacing sensors with Intel Edison

Earlier in this chapter, we had a brief idea about sensors. Now we will see how to interface these sensors with the Intel Edison. Let us consider the use of the temperature sensor. As already mentioned, most of the sensors have a three or four-pin configuration:

- Vcc
- Ground
- Signal

If you have a look at the Edison board, the board will have analog pins. Ideally, if the sensor returns an analog value, then it goes in the analog pin. It is similar for digital output: we prefer the use of digital pins. Let us look at the following example, where we will interface a temperature sensor. Normally, a typical temperature sensor has three pins. The configuration is the same as previously. However, sometimes due to board compatibility issues, it may come with a four-pin configuration, but in that case one of the pins is not used.

In this example, we are going to use a Grove temperature sensor module:

Grove temperature sensor module

The preceding image is a temperature sensor. You may notice that it has four-pins, designated as **Vcc**, **Gnd**, **Sig**, and **NC**. In order to connect with your Edison, follow the following circuit diagram:

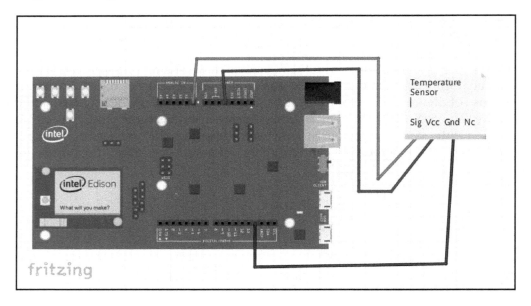

Circuit diagram for basic temperature data

In this circuit diagram, it is noticed that the NC pin is not connected. Thus, only three-pins, **Vcc**, **Gnd**, and **Sig**, are connected. Now, once you are done with the connections, we need to write some algorithms for reading the data. The standard procedure is to search for the datasheet of the sensor. Normally, we also get mathematical equations for the sensor to obtain the desired parameters.

For the Grove temperature sensor module, our first target is to obtain some data from the manufacturer's website. Normally, these sensors calculate temperature based on the change in resistance:

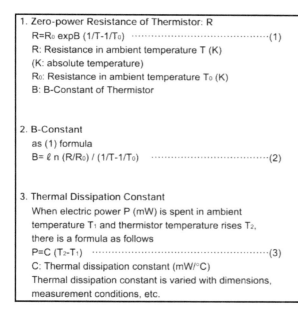
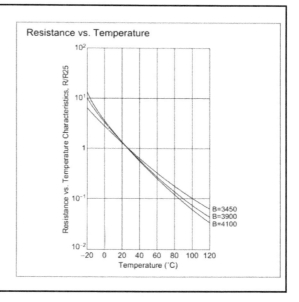

1. Zero-power Resistance of Thermistor: R

$R = R_0 \, expB \, (1/T - 1/T_0)$ ....(1)

R: Resistance in ambient temperature T (K)

(K: absolute temperature)

R_0: Resistance in ambient temperature T_0 (K)

B: B-Constant of Thermistor

2. B-Constant

as (1) formula

$B = \ell n \, (R/R_0) \, / \, (1/T - 1/T_0)$ ....(2)

3. Thermal Dissipation Constant

When electric power P (mW) is spent in ambient temperature T_1 and thermistor temperature rises T_2, there is a formula as follows

$P = C \, (T_2 - T_1)$ ....(3)

C: Thermal dissipation constant (mW/°C)

Thermal dissipation constant is varied with dimensions, measurement conditions, etc.

Resistance vs. Temperature

Algorithm for temperature calculation. Picture credits: http://wiki.seeed.cc/Grove-Temperature_Sensor_V1.2/

The final formulae will be as follows:

B=4275

R0=100000

*R= 1023.0/a-1.0 R= 100000*R*

Final Temperature = *1.0/(log(R/100000.0)/B+1/298.15)-273.15*

The preceding deduction was obtained from http://wiki.seeed.cc/Grove-Temperature_Sensor_V1.2/.

When converting the preceding deduction into code for the Arduino IDE to deploy into the Edison, the result will be as follows:

```
#include <math.h>


constint B=4275; constint R0 = 100000; constint tempPin = A0; void setup()


  {

    Serial.begin(9600);

  }

void loop()

  {

    int a = analogRead(tempPin ); float R = 1023.0/((float)a)-1.0;
    R = 100000.0*R;

    float temperature=1.0/(log(R/100000.0)/B+1/298.15)-273.15;
    Serial.print("temperature = "); Serial.println(temperature);
    delay(500);

  }
```

Explanation of the code

Let's discuss the steps that we need to follow:

1. Initially we declare the values of B and R0. These values are obtained from the datasheet, as shown in the algorithm.

2. Next, we declare the analog pin, tempPin, which will be used.

3. In the setup() function, we just perform the Serial.begin(9600) operation. The thing is that we don't set the tempPin to be in input mode because of the fact that analog pins by default are in input mode.

4. Next in the loop, we implement the calculations performed earlier in code and display it on the serial monitor.

5. To access the serial monitor, press the top right corner button. After the serial windows opens, you will see your current room temperature:

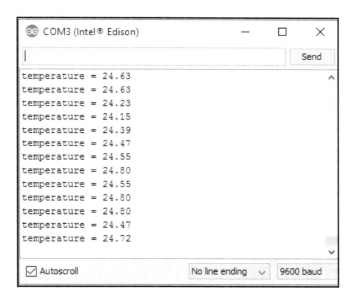

Serial window. Note the baud rate and the temperature readings

From the preceding screenshot, you can note the temperature reading, which is pretty close to the original temperature. Depending on the sensitivity, the temperature may vary.

Now that we know how to access temperature sensor readings, we can move forward with the integration of other sensors for our weather station project.

Connecting the device and uploading to the cloud (dweet.io)

Now that we know how to read the data from the sensors, our next target is to select an IoT platform where we'll upload data. We had a short discussion before about IoT platforms.

In this section, we will be dealing with `dweet.io`. This IoT platform is extremely simple to implement. We will be using the Node.js SDK for `dweet.io`. Before going into the details, let's have a look at the platform. Our target is to push the temperature data we obtained into the platform and display it on a dashboard.

`dweet.io` is a simple publishing and subscribing service for machines (sensors, robots, devices, and so on). It's like Twitter, but for things. Each thing is assigned a unique name and through REST services, we access, and update them. One thing to be noted is that the things we create are public. In order to create a private thing, we need to pay. Here we'll cover only the public aspect of it:

> For details on `dweet.io`, refer to the following link:
> `https://dweet.io/faq`

1. This will be one of the first steps for our weather station, as temperature is an integral part of it:

Dweet.io screenshot

2. Now click on the **Play** tab to enter the creation area:

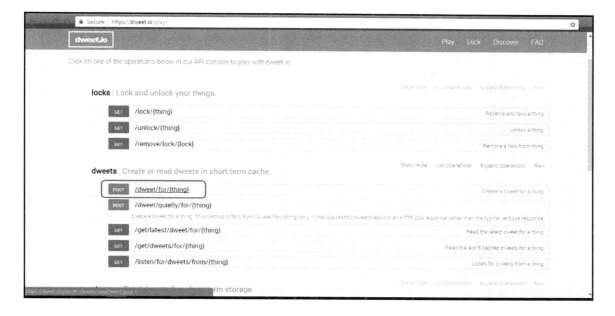

Play—dweet.io

3. Next, we are going to provide a name to our thing:

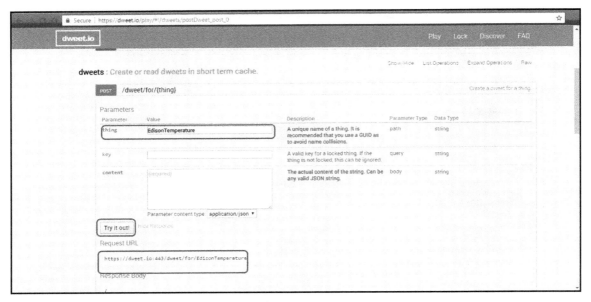

dweet.io—Create a thing

4. Name your thing. The name has to be unique.
5. After clicking on the **Try It Out!** button, you will be getting a request URL.
6. Now, click on `https://dweet.io/follow/EdisonTemperature` to browse to your public thing page. Once the Edison is connected, we'll receive data here. So, for now, let's connect a temperature sensor to the Edison and connect it to your PC.
7. In this mini project, we are going to write a Node.js program that will access the data from the sensor and upload it to here in the preceding link. Also, instead of writing the code to the Intel XDK, we'll write the code in Notepad++ and transfer it using the FileZilla client.

8. But again, we don't have access to the libraries yet. So we need to add those libraries to our Edison device. For that, fire up your PuTTY Terminal and log in to your Edison:

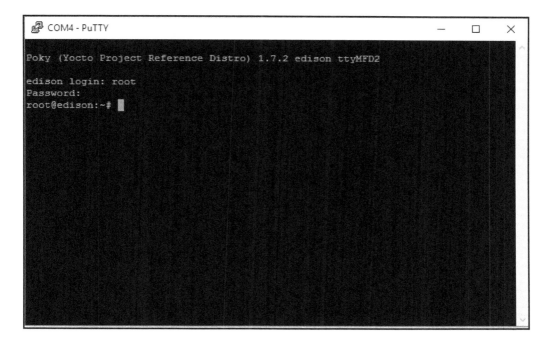

PuTTY console-setup—1

9. Next, check whether your device is connected to your Wi-Fi network using the `ifconfig` command:

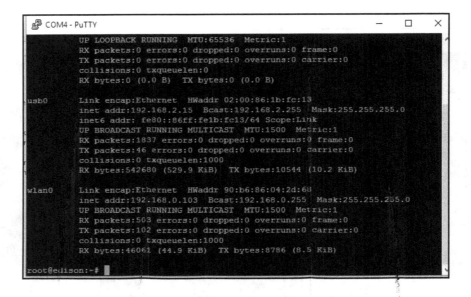

PuTTYconsole-setup—2

10. Next, we will install the Node.js sdk for `dweet.io` on our device. Execute the following command:

```
npm install node-dweetio -save
```

It may take some time for the modules to be installed depending on the speed of your Internet connection. The warnings may be ignored.

11. Once installed, we'll write the Node.js script for the Edison. Now open your editor and type in the following code:

```
function dweetSend()
  {
    vardweetClient = require("node-dweetio"); vardweetio =
    newdweetClient();
    var mraa= require('mraa'); var B=4275;
    var R0=100000;
    var tempPin=new mraa.Aio(0); var a=tempPin.read();
    var R=1023/a-1; R=100000*R;
```

```
var temperature=1/(Math.log(R/100000)/B+1/298.15)-273.15;
temperature = +temperature.toFixed(2);
dweetio.dweet_for("WeatherStation",
{Temperature:temperature}, function(err, dweet)
  {
    console.log(temperature); console.log(dweet.thing); //
    "my-thing"
    console.log(dweet.content); // The content
    of the dweet
    console.log(dweet.created); // The create
    date of the dweet
  });
setTimeout(dweetSend,10000);
}
dweetSend();
```

To explain the code, let's break it into parts. As you can see, we have a main function named dweetSend, which repeats itself after every 10 seconds. Initially, we need to get data from the temperature sensor. We've connected the temperature sensor to Analog pin 3. If you have a close look at the code that was written in Arduino IDE, then you will find lots of similarities:

```
vardweetClient = require("node-dweetio");
vardweetio = new dweetClient();
```

In these lines, we import the node - dweetio library. The next piece of code is similar to the Arduino IDE, where we read the raw value of the analog read and perform required calculations:

```
var mraa= require('mraa');
var B=4275; varR0=100000;
var tempPin=new mraa.Aio(0);
var a=tempPin.read();
var R=1023/a-1;
R=100000*R;
var temperature=1/(Math.log(R/100000)/B+1/298.15)-273.15; temperature =
+temperature.toFixed(2);
```

As stated in the preceding code, we have also rounded the temperature value to two decimal places. Next, we push the value to the dweet.io thing channel. Here, we need to mention that our thing name WeatherStation, followed by the name of the parameter, Temperature, and the variable name, temperature:

```
dweetio.dweet_for("WeatherStation", {Temperature:temperature},
function(err, dweet)
  {
```

```
    console.log(temperature); console.log(dweet.thing); // "my-thing"
    console.log(dweet.content); // The content of the dweet
    console.log(dweet.created); // The create date of the dweet
});
```

That's how the entire workflow is. So to summarize:

1. Import required libraries.
2. Based on the circuit, read the raw values from the used pin and process the value to get the desired output.
3. Push the value to dweet.io.

Save this code with the name dweetEdison.js and to run this code, type in the following command:

node dweetEdison.js

Once you run the code by executing the preceding statement in the PuTTY Terminal, you will see the following output:

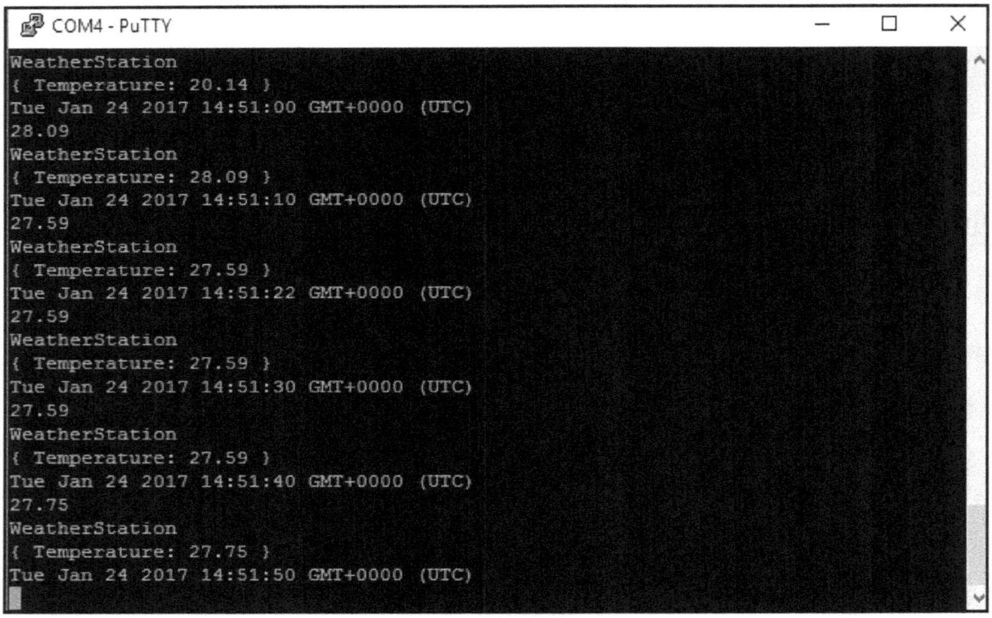

PuTTY console—output

Sometimes, the preceding code may throw an error while importing the mraa library. There are issues with reading analog pin values. This usually happens if you have the mraa library installed from multiple sources. In that case, reflash your device and follow all the steps.

Now, once you see this, then your code is running successfully. Now we will head on to the dweet.io page and see whether our data is really getting uploaded or not. We have named our thing **WeatherStation**. Your thing name will obviously differ since it's unique.

Now browse to the following URL:

```
https://dweet.io/follow/YOUR_THING_NAME
```

You should have a plot or a visual like this:

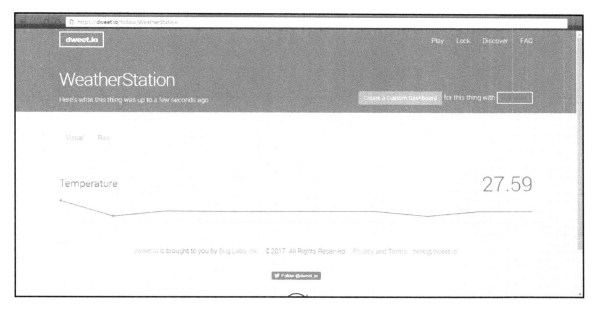

Temperature plot—dweet.io

The plot seems incomplete unless and until we have a gauge. There is something called `freeboard.io`. It's also free, but again it is public. So, browse to `freeboard.io` and log in or sign up if you don't have an account:

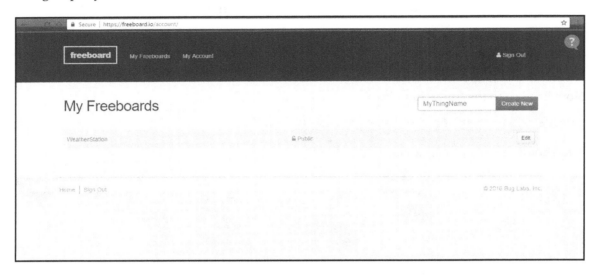

freeboard.io—account page

Provide a name and click on **Create New**. Here we have provided **WeatherStation**. Once you create the board, you will be automatically redirected to the board design page. Next, under **DATASOURCES**, click on **ADD**:

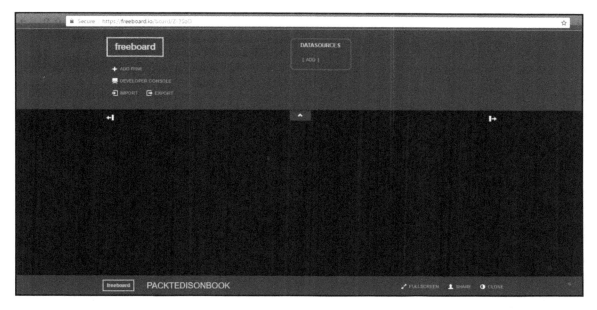

Freeboard screenshot

Now, once you click on **ADD**, you have to select a **TYPE**. Select **Dweet.io**. Under **NAME**, provide any name. For simplicity, make the name and the thing name the same. Since your thing in `dweet.io` is public, we don't need to provide a key. Click on **SAVE** to proceed:

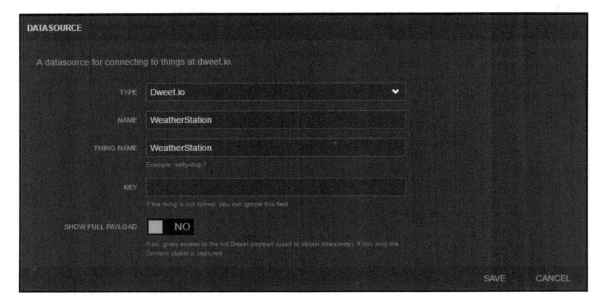

Dweet.io datasource

Once done, we need to add a pane for our incoming data. For that, click on **ADD PANE** and proceed to add a new gauge. Fill in the details as shown in the following screenshot. Note that under the value field, we've written

`datasources["WeatherStation"]["Temperature"]`. So here, `WeatherStation` is your thing name, which is followed by the parameter name that we want to display. Similarly, the project that will be discussed in the upcoming section of this chapter will have other parameters too:

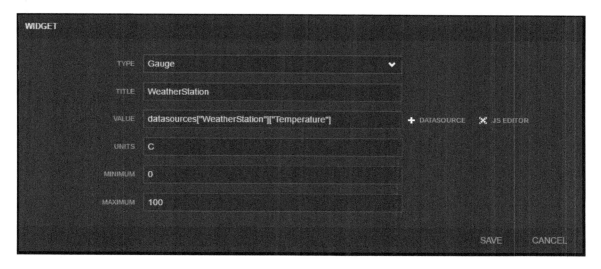

Freeboard

Click on **SAVE** and you should have your gauge displayed on your board home page.

This gauge won't show any value as of now since your code is not running. So go back to your PuTTY console and run your Node.js code. Once your code is up and running, then the gauge should represent your incoming data, as shown in the following screenshot:

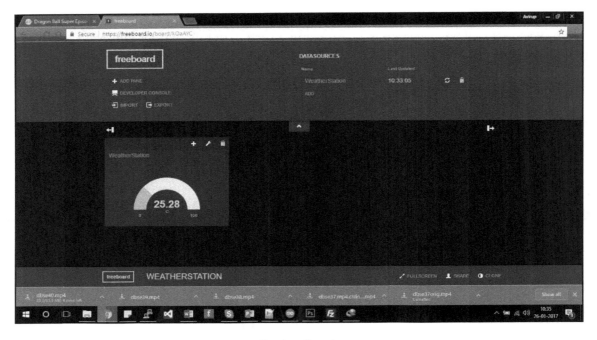

Gauge in working mode

Thus we have deployed a very simple use case of uploading data from the Edison to `dweet.io` and displaying it on `freeboard.io`. In the next section, we'll deal with a live IoT project where we'll deal with more than one parameter.

Live use case of an IoT project - mini weather station

In the previous section, we have seen the use case of uploading temperature data only. Now, with mini weather station, we are going to deal with three parameters:

- Temperature
- Pollution level
- Sound level

All three parameters will be uploaded to the cloud and displayed live on the dashboard. The project will be divided into parts, which will make it easier to understand, and we will also tackle other complex projects:

- Architecture of the system
- Sensor and hardware components and detailed circuit diagram
- Locally acquiring data and displaying it on the console
- Uploading the data to the cloud
- Visualizing the data

Most details were shown previously using a temperature sensor, so we won't go into much depth, but we will cover all aspects.

Architecture of the system

The architecture of the system will be very much similar to the generic architecture of the system discussed at the beginning of the chapter, with the components of actuators and alerts missing:

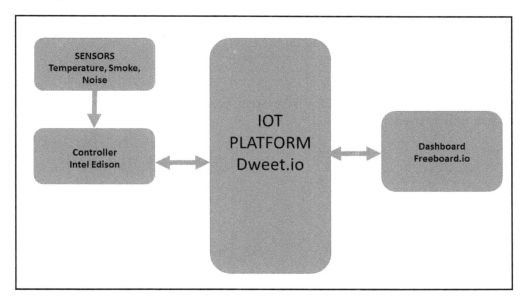

Architecture of the weather station

As you can see in the preceding figure, we have three sensors (temperature, smoke, and noise) that supply raw data to the controller, that is Intel Edison, and the controller processes the raw data to upload to the IoT platform of `dweet.io`, which is ultimately displayed on the dashboard on `freeboard.io`. For this project, we are not interested in alerts and actuators, so those aren't included.

Hardware components and detailed circuit diagram

For the project to be realized, we will deal with the following hardware components:

- Grove temperature sensor
- Grove sound sensor
- MQ2 smoke sensor

All of the preceding three sensors have more or less a similar configuration of pins, so connections are pretty simple. As already explained, we're going to use the analog pins on the Intel Edison for the input from the sensors.

 The MQ2 sensor returns raw values. Here, we'll use the raw values and calibrate the sensor based on the level of smoke.

The following is the circuit diagram for the system. Please note that we've only mentioned the pins that will be used:

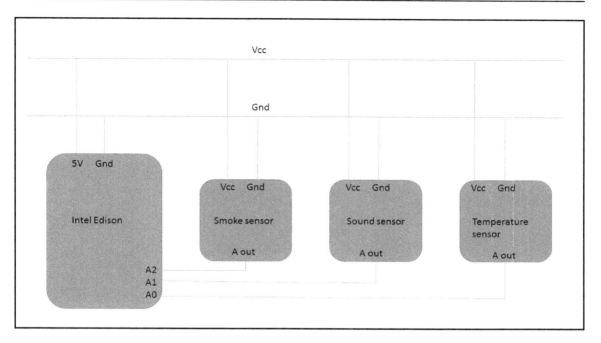

Circuit diagram - weather station

We've made a common **Vcc** and **Gnd** connection where all the sensors are hooked up. Next, we need three analog pins. These pins are connected to the analog output pins of the sensors. It should be noted that some sensors may have two output pins, one of which one is analog out, while the other is digital out. Since we are interested in analog output, we will not be using the digital output pins. In this project, the MQ2 sensor (the smoke sensor) has such a configuration.

For calibrating the **Smoke sensor**, we need the presence of some smoke that the sensor is sensitive to. Although we'll be pushing raw values, for local alarms the thresholding technique may be used.

Connect your sensor and the Edison device based on the preceding circuit diagram. Once connected, you are good to go with the code.

Code for weather station stage 1, acquiring data from all the sensors and displaying it in the console

Before moving on to the code, let's have a look at the algorithm first:

1. Import libraries.
2. Initialize input pins.
3. Read raw values.
4. Process the values.
5. Display it to the console.

This is similar to the earlier example, the only difference being that we're going to use three sensors. In this case, the code will be written in Node.js, since at a later stage we'll be pushing it to the cloud, that is, dweet.io:

```
functiondisplayRes()

{

//Import mraa

var mraa= require('mraa'); var B=4275;
var R0=100000;

//Temperature pin

var tempPin=new mraa.Aio(0);

//Sound pin

varsoundPin= new mraa.Aio(1);

//Smoke pin

varpolPin= new mraa.Aio(2);

//Processing of temperature var a=tempPin.read();
var R=1023/a-1; R=100000*R;
var temperature=1/(Math.log(R/100000)/B+1/298.15)-273.15; temperature =
+temperature.toFixed(2);
//Smoke read

varsmokeValue= polPin.read();
```

```
//Sound read

varsndRead= soundPin.read();


console.log("Temperature=   ",temperature);console.log("Soundlevel=
",sndRead);console.log("Smoke level= ", smokeValue);
setTimeout(displayRes,500);
}

displayRes();
```

Before explaining the code, processing has only been performed on the temperature sensor. For the sound level, we'll send raw values because for conversion into decibels, which is a relative quantity, we need to access the sound pressure of two instances. So we will restrict ourselves to raw values. However, we can certainly find a threshold value of the raw readings and use the threshold to invoke an action, such as turning on an LED or sounding a buzzer.

Now, let's have a close look at the code. Most of the code is similar to that of the temperature module. We've added a few more lines for smoke and sound detection:

```
//Sound pin
varsoundPin= new mraa.Aio(1);
//Smoke pin
varpolPin= new mraa.Aio(2);
```

In the preceding lines, we declared which analog pins are used for sound sensor input and smoke sensor input. In the following lines we will read the values:

```
//Smoke read
varsmokeValue= polPin.read();
//Sound read
varsndRead= soundPin.read();
```

Ultimately, we display the captured values using the console.

 When dealing with sensors such as smoke and sound sensors, we may not get the standard unit values directly. In these cases, we need to manually calibrate the sensors to simulate an environment of known values. Suppose in case of a smoke sensor, we set up an environment where we know what the value is, then by varying the potentiometer we change the value of the variable resistance, thus calibrating the sensor. That's one of the many standard procedures available. Calibration will be dealt with in detail when we cover the robotics module.

When you run the preceding code in the console, you will get the output from all the sensors. Try to increase smoke around the smoke sensor or speak loudly in front of the sound sensor to increase the value, or keep the temperature sensor near your laptop vent to get a higher reading. The following is the screenshot of the values obtained from the sensors:

Sensor reading output

Once you obtain the readings, we can push them to the cloud and display them on the dashboard.

Here, if you notice that you are not getting correct readings, then you need to adjust the potentiometer available on the sensor to calibrate it manually. For uploading it to the cloud, we need to impart some changes in the code. Refer to the following code for pushing all the three data obtained to dweet.io:

```
function dweetSend()
    {
        vardweetClient = require("node-dweetio"); vardweetio = new
        dweetClient();
//Import mraa
        var mraa= require('mraa'); var B=4275;
        var R0=100000;
```

```
//Temperature pin
    var tempPin=new mraa.Aio(0);
//Sound pin
    varsoundPin= new mraa.Aio(1);
//Smoke pin
    varpolPin= new mraa.Aio(2);
//Processing of temperature var a=tempPin.read();
    var R=1023/a-1; R=100000*R;
    var temperature=1/(Math.log(R/100000)/B+1/298.15)-273.15;
    temperature = +temperature.toFixed(2);
//Smoke read
    varsmokeValue= polPin.read();
//Sound read
    varsndRead= soundPin.read();
    dweetio.dweet_for("WeatherStation",
      {Temperature:temperature, SmokeLevel:smokeValue,
      SoundLevel:sndRead}, function(err, dweet)
        {
          console.log(dweet.thing); // "my-thing"
          console.log(dweet.content); // The content
          of the dweet
          console.log(dweet.created); // The create
          date of the dweet
        });
      setTimeout(dweetSend,10000);
    }
dweetSend();
```

In the preceding code, again you will find lots of similarities with the code for `temperature`. Here, we have performed three read operations and we've sent all the three values respective to the parameter it represents. It's evident from the following line:

```
dweetio.dweet_for("WeatherStation", {Temperature:temperature,
SmokeLevel:smokeValue, SoundLevel:sndRead}, function(err, dweet)
```

Transfer the code by following a similar process that was discussed before, using FileZilla, and execute it using the `node` command:

```
COM4 - PuTTY                                                          —   □   ×
{ Temperature: 20.86, SmokeLevel: 284, SoundLevel: 5 }
Fri Jan 27 2017 14:30:04 GMT+0000 (UTC)
WeatherStation
{ Temperature: 20.62, SmokeLevel: 282, SoundLevel: 16 }
Fri Jan 27 2017 14:30:24 GMT+0000 (UTC)
^Croot@edison:~# node dweetEdisonTest.js
WeatherStation
{ Temperature: 20.86, SmokeLevel: 283, SoundLevel: 0 }
Fri Jan 27 2017 14:30:53 GMT+0000 (UTC)
WeatherStation
{ Temperature: 20.86, SmokeLevel: 287, SoundLevel: 0 }
Fri Jan 27 2017 14:31:11 GMT+0000 (UTC)
WeatherStation
{ Temperature: 20.62, SmokeLevel: 283, SoundLevel: 0 }
Fri Jan 27 2017 14:31:31 GMT+0000 (UTC)
WeatherStation
{ Temperature: 20.54, SmokeLevel: 281, SoundLevel: 0 }
Fri Jan 27 2017 14:31:51 GMT+0000 (UTC)
WeatherStation
{ Temperature: 19.82, SmokeLevel: 279, SoundLevel: 0 }
Fri Jan 27 2017 14:32:11 GMT+0000 (UTC)
WeatherStation
{ Temperature: 19.66, SmokeLevel: 374, SoundLevel: 34 }
Fri Jan 27 2017 14:32:31 GMT+0000 (UTC)
```

PuTTY Terminal

Having a look at the preceding screenshot, it's clear that values are being sent. Now, please note the sound and smoke values. Initially, music was being played, so we got sound values in the range of 20-70. For the smoke sensor, the standard value is around 250-300. In the last reading, I applied some smoke and it shot to **374**. Now browse to your `dweet.io` portal and you will notice the values being updated live:

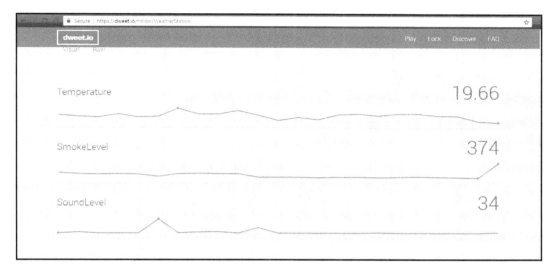

Dweet.io screenshot for live data

Once we have things set up on this side, we'll add two more gauges to `freeboard.io` for visualizations. Log on to `freeboard.io` and follow the method as explained before for addition of gauges. Be specific when matching the **DATASOURCES** where you need to specify the parameter:

Freeboard.io final layout

Once this is done, well, you will have your own weather station ready, and up and running. Once we understand the concepts, it's extremely easy to realize these mini projects.

Open-ended task for the reader

Since you have got a fair idea of how to realize small IoT projects, we'll provide you a simple open-ended task for you to implement.

We've used `dweet.io` and `freeboard.io`. Besides this, there are several other IoT platforms, such as Azure, IBM Bluemix, and Thingspeak. Your task is to simulate a similar thing on other IoT platforms. Azure and IBM requires you to sign up by the use of debit or credit card, but for the first month it's free and it can be used for IoT, while, on the other hand, Thingspeak is totally free and it has support for Matlab.

Summary

In this chapter, we've learned about the architecture of IoT systems, how to tackle problem statements, and we also learned about sensors and interfacing them. Then we moved on to the IoT platforms and visualized some temperature data. At the end, we built a mini weather station that measures temperature, sound level, and pollution level. We also covered a very brief discussion of the calibration of sensors.

In Chapter 3, *Intel Edison and IoT (Home Automation)*, we'll be dealing with communication from a platform to the controller, that is, controlling devices using the Internet. Then we'll engage ourselves in a home automation project where we'll be controlling two devices using an Android and a WPF application.

3
Intel Edison and IoT (Home Automation)

In Chapter 2, *Weather Station (IoT)*, we dealt with transferring data from Edison to the cloud platform. Here, in this chapter, we'll be doing just the opposite. We'll be controlling devices using the Internet. When we talk about IoT, the first thing that usually comes to mind is home automation. Home automation is basically controlling and monitoring home electrical appliances using an interface, which may be a mobile application, a web interface, a wall touch unit, or more simply, your own voice. So, here in this chapter, we'll be dealing with the various concepts of home automation using the MQTT protocol; then, we'll be controlling an electrical load with an Android application and a **Windows Presentation Foundation (WPF)** application using the MQTT protocol. Some of the topics that we will discuss are:

- The various concepts of controlling devices using the Internet MQTT protocol

- Using Edison to push data and get data using the MQTT protocol

- LED control using the MQTT protocol

- Home automation use cases using the MQTT protocol

- The controller application in Android (MyMqtt) and in WPF (to be developed)

This chapter will use a companion application named MyMqtt, which can be downloaded from the Play Store. Credit goes to the developer (Instant Solutions) for developing the application and uploading it to the Play Store for free. MyMqtt can be found here:
https://play.google.com/store/apps/details?id=at.tripwire.mqtt.client&hl=en

We are going to develop our own controller for PC as a WPF application that will implement the protocol and control your Edison.

 To develop the WPF application, we are going to use Microsoft Visual Studio. You can download it at `https://msdn.microsoft.com`.

Controlling devices using the Internet - concepts

When it comes to control devices using the Internet, some key factors come to play. Firstly, is the technique to be used. There are lot of techniques in this field. A quick workaround is the use of REST services, such as HTTP GET requests, where we get data from an existing database.

Some of the workarounds are discussed here.

REST services

One of the most commonly-used techniques for obtaining the desired data is by an HTTP GET call. Most of the IoT platforms that exist in the market have REST APIs exposed. There, we can send values from the device to the platform using an HTTP POST request, and at the same time get data by an HTTP GET request. Infact, in Chapter 2, *Weather Station (IoT)*, where we used dweet.io to send data from a device, we used an SDK. Internally, the SDK also performs a similar HTTP POST call to send in data.

Instructions or alerts (present on most IoT platforms)

In certain IoT platforms, we have certain ready-made solutions where we just need to call a certain web service and the connection is established. Internally, it may use REST APIs, but for the benefit of the user, they have come up with their own SDK where we implement.

Internally, a platform may follow either a REST call, MQTT, or Web Sockets. However, we just use an SDK where we don't implement it directly, and by using the platform's SDK, we are able to establish a connection. It is entirely platform-specific. Here, we are discussing one of the workarounds,where we use the MQTT protocol to control our devices directly without the use of any IoT platforms.

Architecture

In a typical system, the IoT platform acts as a bridge between the user and the protocols to the controller, as shown in the following figure:

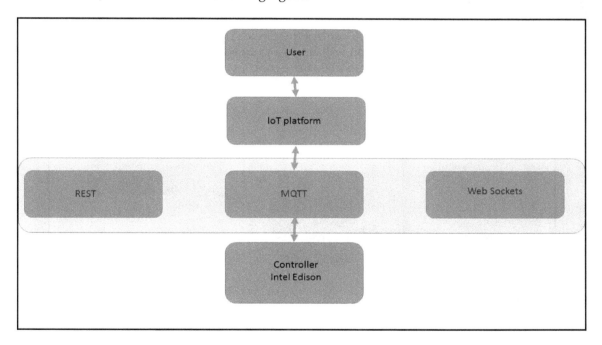

Architecture of the IoT system for controlling devices

The preceding image depicts a typical workflow or architecture of controlling devices using the Internet. It is to be noted that the user may directly control the controller without the use of an IoT platform, as we do here. However, normally a user will use the IoT platform, which also provides more enhanced security. The user may use any web interface, mobile application, or a wall control unit to control the device using any standard protocol. Here in the image, only REST, MQTT, and Web Sockets are included. However, there are more protocols that can be used, such as the AMQP protocol, the MODBUS protocol, and so on.

The choice of the protocol depends mainly on how sensitive the system is and how stable the system needs to be.

MQTT protocol overview

The MQTT protocol is based on the publish-subscribe architecture. It's a very lightweight protocol, where message exchange happens asynchronously. The main usage of the MQTT protocol is in places of low bandwidth and low processing power. A small code footprint is required for establishing an MQTT connection. Every communication in the MQTT protocol happens through a medium called a broker. The broker is either subscribed or published. If you want the data to flow from Edison to a server, then you publish the data via the broker. A dashboard or an application subscribes to the broker with the channel credentials and provides the data. Similarly, when we control the device from any application, Edison will act as a subscriber and our application will act as a publisher. That's how the entire system works out. The following screenshot explains the concept:

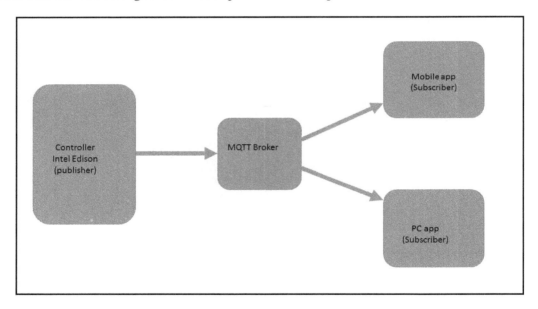

Overflow where Edison acts as a publisher

In the preceding screenshot, we see Edison acting as a publisher. This is one type of use case, where we need to send data from Edison, as with a similar example shown in Chapter 2, *Weather Station (IoT)*. The application will get the data and act as a publisher. The following screenshot depicts the use case that will be used in this chapter: the use of Edison as a subscriber:

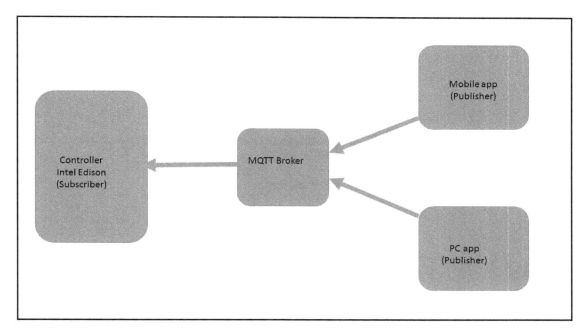

Overflow where Edison acts as a subscriber

In the preceding case, we have some controls on the application. These controls send signals to Edison via the MQTT broker. Now, in this case, the application will act as a publisher and Edison acts as a subscriber.

It is to be noted that in a single system, you can make the endpoint (device or application) act both as a publisher as well as a subscriber. This occurs when we want to get some data from the IoT device, such as the Intel Edison, and also control the device in emergency cases. The same may also occur when we need to control the home's electrical appliances, as well as monitor them remotely. Although most systems are deployed based on a closed loop feedback control, there is always room to monitor them remotely, and at the same time have control based on feedback received from the sensors.

To implement the MQTT protocol, we are not going to set our own server but use an existing one. `https://iot.eclipse.org/` has provided a sandbox server which will be used for the upcoming projects. We're just going to set up our broker and then publish and subscribe to the broker. For the Intel Edison side, we are going for Node.js and its related libraries. For the application end, we are going to use an already available application named MyMqtt for Android. If anyone wants to develop his or her own application, then you need to import the `paho` library to set up MQTT. We are also developing a PC application, where we will again use MQTT to communicate.

For details on the eclipse IoT project on MQTT and other standards, please refer to the following link:

`https://iot.eclipse.org/standards/`

In the following section, we'll set up and configure Edison for our project and also set up the development environment for the WPF application.

The paho project can be accessed through this link:

`https://eclipse.org/paho/`

Using Intel Edison to push data by using the MQTT protocol

As previously mentioned, this short section will show users how to push data from Edison to an Android device using the MQTT protocol. The following screenshot depicts the workflow:

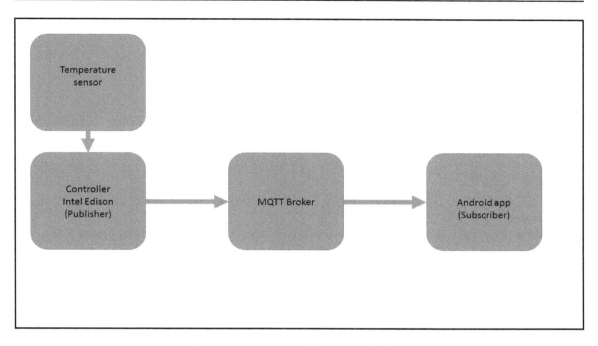

Workflow of pushing data from the Edison to the Android application

From the preceding illustration, it is clear that we first obtain readings from the temperature sensor and then use the MQTT broker to push the readings to the Android application.

Firstly, we are going to connect the temperature sensor to Edison. Make a reference of the circuit from Chapter 2, *Weather Station (IoT)*. After it is connected, fire up your editor to write the following Node.js code:

```
var mraa = require('mraa'); var mqtt = require('mqtt'); var B=4275;

var R0=100000;

var client = mqtt.connect('mqtt://iot.eclipse.org');
function sendData()

{

  var tempPin=new mraa.Aio(0);

//Processing of temperature var a=tempPin.read();

  var R=1023/a-1; R=100000*R;
```

```
var temperature=1/(Math.log(R/100000)/B+1/298.15)-273.15; temperature
= +temperature.toFixed(2);

//Converting type int to type string

  var sendTemp= temperature.toString();

//Publish the processed data client.publish('avirup/temperature',sendTemp);
console.log("Sending data of temperature %d", temperature);
setTimeout(sendData,1000);

}

sendData();
```

The code written here is similar to what we used in the Chapter 2, *Weather Station (IoT)*. Here, the difference is that we are not sending it to dweet.io but to the MQTT broker. We're publishing the data obtained to a particular channel in the MQTT broker.

However, to execute this code, you must have the MQTT dependency installed via npm. Type in the following command in the PuTTY console:

npm install mqtt

This will install the MQTT dependency.

In the preceding code, we initially imported the required libraries or dependency. For this case, we need the mraa and the mqtt libraries:

```
var mraa = require('mraa');
var mqtt = require('mqtt');
```

Then, we need to initialize the analog pin to read the temperature. After that, we convert the raw readings to the standard value.

We declare the client variable, which will handle the MQTT publish operation:

```
var client = mqtt.connect('mqtt://iot.eclipse.org');
```

Here, https://iot.eclipse.org/ is the free broker that we are using.

Next, in the sendData function, the initial temperature processing is computed before the data is published to the channel:

```
client.publish('avirup/temperature',sendTemp);
```

The name of the channel is `avirup/temperature`. Please note the type of `sendTemp`. The initial processed value is obtained in the variable temperature. Here in `client.publish`, the second parameter has to be a string. Thus, we store the temperature value as a string type in `sendTemp`. Finally, we print the temperature into the console.

We have also provided a delay of 1 second. Now run this Node.js file using the `node` command.

The screenshot is as follows:

```
COM4 - PuTTY                                          —    □    ×
root@edison:~# node mqttPublisher.js
Sending data of temperature 25.77
Sending data of temperature 24.88
Sending data of temperature 25.69
Sending data of temperature 24.88
Sending data of temperature 25.45
Sending data of temperature 25.37
Sending data of temperature 25.28
Sending data of temperature 25.61
Sending data of temperature 25.53
```

Output console log

As seen in the preceding screenshot, the log is displayed. Now we need to see this data in the Android MyMqtt application.

While carrying out this mini-project, as well as the later one to be discussed under MQTT, please change the channel name. One of my projects may be live and it could create an issue. One can go for the `NAME_OF_THE_USER/VARIABLE_NAME` convention.

Open up the MyMqtt application in Android and browse to **Settings**. There, in the field of **Broker URL**, insert `iot.eclipse.org`. You will have used this on your Node.js snippet as well:

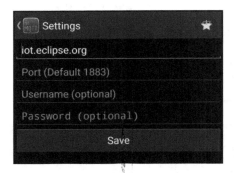

Screenshot of MyMqtt—1

Next, go to the **Subscribe** option and enter your channel name based on your Node.js code. In our case, it was `avirup/temperature`:

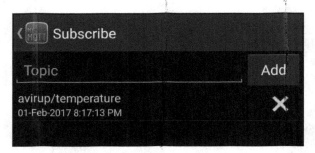

Screenshot of MyMqtt—2

Click on **Add** to add the channel and then finally go to the dashboard to visualize your data:

Screenshot of MyMqtt—3

If your code on the device is running in parallel to this, then you should get live data feed in this dashboard.

So, now you can visualize the data that you are sending from Edison.

Getting data to Edison by using MQTT

We have been talking about home automation controlling electrical loads, but everything has a starting point. The most basic kick-starter is controlling Edison over the Internet—that's what it's all about.

When you have a device that is controllable over the Internet, we recommend controlling the electrical loads. In this other mini-project, we are going to control a simple LED that is already attached to pin 13 of Intel Edison. There is no need for any external hardware for this, as we are using an in-built functionality. Now, open your editor and type in the following code:

```
var mraa = require('mraa'); var mqtt = require('mqtt');

varledPin=new mraa.Gpio(13); ledPin.dir(mraa.DIR_OUT);

var client = mqtt.connect('mqtt://iot.eclipse.org');
client.subscribe('avirup/control/#')
client.handleMessage=function(packet,callback)

{

  var payload = packet.payload.toString() console.log(payload);
  if(payload=='ON')

  ledPin.write(1); if(payload=='OFF') ledPin.write(0);
  callback();

}
```

The preceding code will subscribe to the channel in the broker and wait for incoming signals.

Initially, we've declared the GPIO pin 13 as the output mode because the onboard LED is connected to this pin:

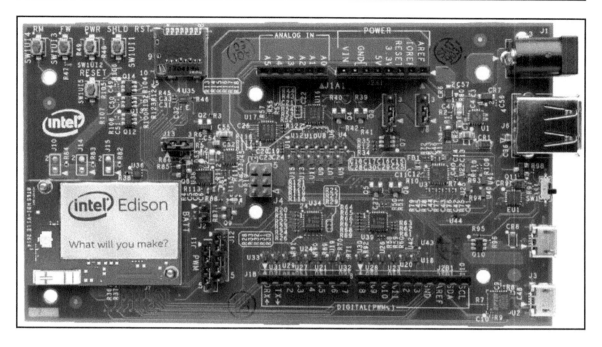

Onboard LED location

The location of the onboard LED is shown in the preceding image.

On having a close look at the code, we see that it initially imports the library and then sets the GPIO pin configuration. Then, we use a variable client to initiate the MQTT connection to the broker.

After that, we move on to subscribe our device to the channel, which in this case is named as `avirup/control/#`.

We have an event handler, `handleMessage()`. This event handler will deal with incoming messages. The incoming message will be stored in the packet variable. We've also implemented a callback method, `callback()`, which needs to be called from `handleMessage()`.

This enables us to receive multiple messages. Also note that, unlike other Node.js snippets, we haven't implemented any loop. The functionality is actually handled by the `callback()` method.

Finally, inside the function we obtain the payload, which is the message. It is then converted to a string and then condition checking is performed. We also print the value received to the console.

Now push this code to your Edison using FileZilla and run the code.

Once you run the code, you won't see anything in the console. The reason behind that is there is no message. Now, go to the Android application, MyMqtt, and browse to the **Publish** section of the application.

We need to insert the channel name here. In this case, it is `avirup/control`:

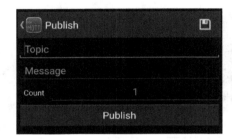

Publish MyMqtt

In the **Topic** section, enter the channel name, and in the **Message** section enter the message to be sent to Edison.

Now, in parallel, run your Node.js code.

Once your code is up and running, we will send a message. Type `ON` in the **Message** field and click **Publish**:

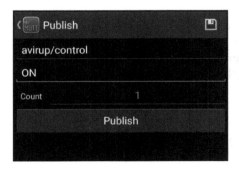

Send control signals

Once you have published from the application, it should be reflected on the PuTTY console:

Message send and receive—MQTT

Now you should see that the LED is turned on.

Similarly, send a message, OFF, to turn off the onboard LED:

Message send and receive. The LED should turn off

It's also worth noting that this will work even if Edison and the device aren't connected to the same network.

Now you can control your Intel Edison with your Android application. Virtually speaking, you can now control your home. In the following section, we'll deep dive into the home automation scenario and also develop a WPF application to control.

Home automation using Intel Edison, MQTT, Android, and WPF

Until now we have learned about the MQTT protocol and how to subscribe and publish data, both using the application and Edison. Now we will be dealing with a real use case where we'll control an electrical load using Intel Edison, which again will be controlled by the Internet. Here is a quick introduction about what we will be dealing with:

- Hardware components and circuits
- Developing a WPF application to control Intel Edison
- Using MQTT to stitch everything together

Since we've already seen how to control Edison using an Android application, this section won't concentrate on that; instead, it will mainly deal with the WPF application. This is just to give you a brief idea about how a PC can control IoT devices, not only in home automation, but also in various other use cases, both in simple proof of concept scenarios to industry standard solutions.

Hardware components and circuit

When we are dealing with electrical load, we simply cannot directly connect it to Edison or any other boards, as it will end up frying. For dealing with these loads, an interfacing circuit is used called a relay. A relay in its crude form is a series of electromechanical switches. They operate on a DC voltage and control AC sources. Components that will be used are listed as follows:

- Intel Edison
- 5V relay module
- Electric bulb wires

Before going into the circuitry, we'll discuss relays first:

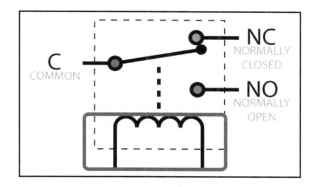

Relay schematics. Picture credits: http://www.phidgets.com/docs/3051_User_Guide

The red rectangular area represents the electromagnet. We excite the electromagnet with a DC voltage, and that triggers the mechanical switch. Having a closer look at the preceding image, we can see three ports where the AC load is connected: common, normally closed, and normally open. In default conditions, that is when the electromagnet is not excited, and the common and normally closed ports are connected. What we are interested in for now is the normally open port.

The image of the relay used is shown as follows:

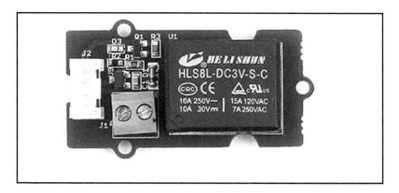

Relay unit. Picture credits: Seed Studio

The electrical load will have a live and neutral wire. Connect either one according to the following circuit:

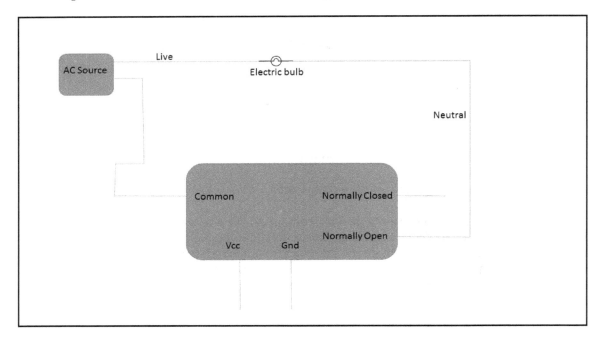

Basic relay connection

With reference to the preceding figure, **Vcc** and **Gnd** are connected to the controller. The AC source connects one end of the electrical load directly, while the other is via the relay. A part of it connects the common port, while the other may be in **normally closed** (**NC**) or **normally open** (**NO**). When you have the other end of the electrical load connected to the NC port, then by default without excitation of the electromagnet, the circuit is complete. Since we don't want the bulb to be operating when the electromagnet isn't excited, connect it to the **NO** port, rather than **NC**. Thus, when the electromagnet is operating by applying voltage on **Vcc** and **Gnd** as ground, the mechanical switch flips to the **NO** position, thus connecting it with the common port.

The whole idea behind the operation of a relay is the use of electromechanical switches to complete a circuit. However, it is worth noting that not all relays operate on the same principle; some relays use solid state devices to operate.

Solid State Relays (**SSRs**) don't have any movable parts unlike that of electromechanical relays. SSRs uses photo-couplers to isolate the input and the output. They change electrical signals to optical signals, which propagates through space and thus isolates the entire circuit. The coupler on the receiving end is connected to any switching device, such as a MOSFET, to perform the switching action.

There are some advantages of using SSRs over electromechanical relays. They are as follows:

- They provide high speed, high frequency switching operations
- There is failure of contact points
- They generate minimal noise
- They don't generate operation noise

Although we will use electromechanical relays for now, if the use case deals with high frequency switching, then it's better to go with SSRs. It is also to be noted that when exposed to long usage, SSRs are known to heat up.

Final circuit

The entire connection is shown in the following figure:

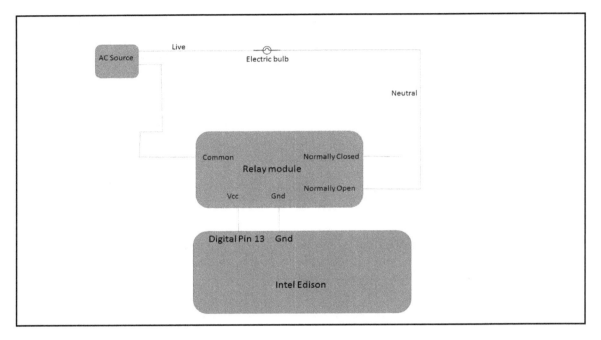

Circuit diagram for home automation project

The circuit adds Intel Edison because the relay circuit will be controlled by the controller. The relay here just acts as an interfacing unit to the AC load.

 While the relay is being operated, please do not touch the underside of it or you may get an AC electric shock, which can be dangerous.

To test whether the circuit is working or not, try out a simple program using the Arduino IDE:

```
#define RELAY_PIN 13 void setup()
{
  pinMode(RELAY_PIN,OUTPUT); //Set relay pin to output
}
void loop
{
  digitalWrite(RELAY_PIN, HIGH); //Set relay to on position
}
```

The code should switch the position of the switch from the NC position to the NO position, thus completing the circuit, leading your bulb to glow. Don't forget to switch on the AC power supply.

Once you have the final circuit ready, we'll move forward with the development of the WPF application, which will control Edison.

Android application for controlling Intel Edison using MQTT

In the previous section, we saw how an Android application can be used to subscribe and publish to a channel using a broker. Here, in this section, we'll develop our own Android application for controlling the device using MQTT. The section won't concentrate on the set up of the Android, but will concentrate on the development side of it. We're going to use the Android Studio IDE for the development of the application. Make sure it's configured with all the latest SDKs.

Open your **Android Studio**:

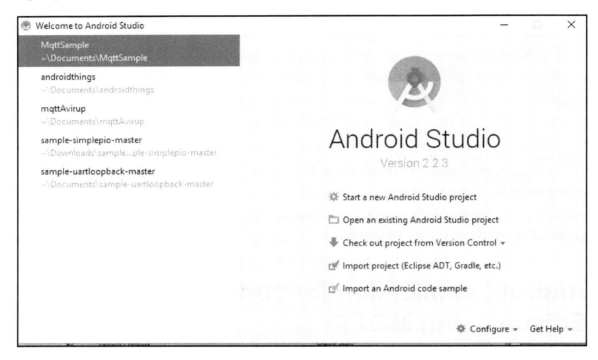

Android Studio—1

Now, select **Start a new Android Studio project**:

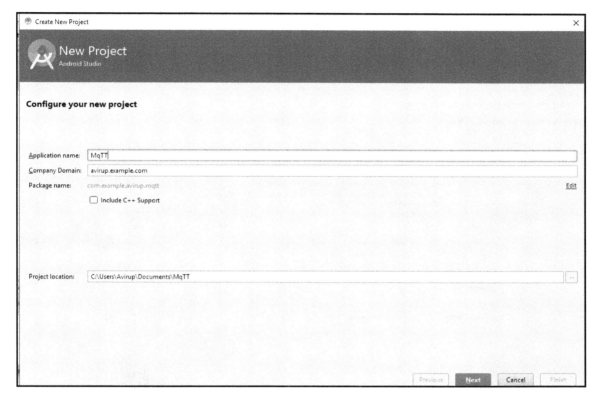

Android Studio—set up application name

Enter a name for your application; here, we've entered MQTT. Click on **Next** to continue:

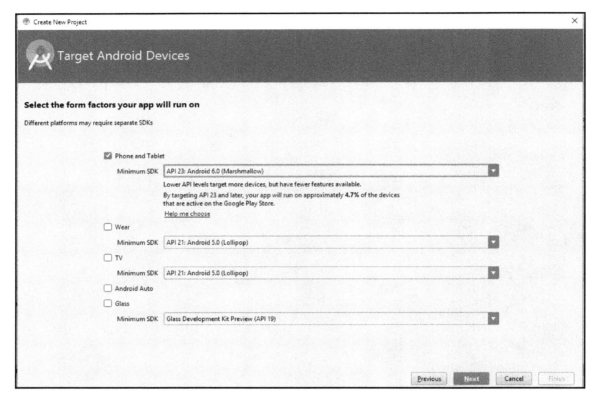

Android Studio: set API level

Now select the **Minimum SDK** version. Select **API 23: Android 6.0 (Marshmallow)**. Now let's select the type of activity:

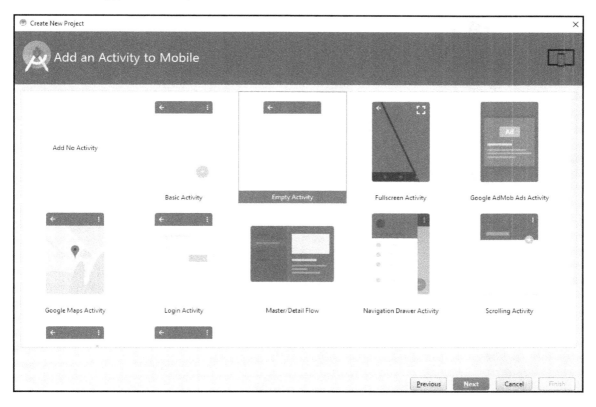

Set activity

Select **Empty Activity** and click on **Next**:

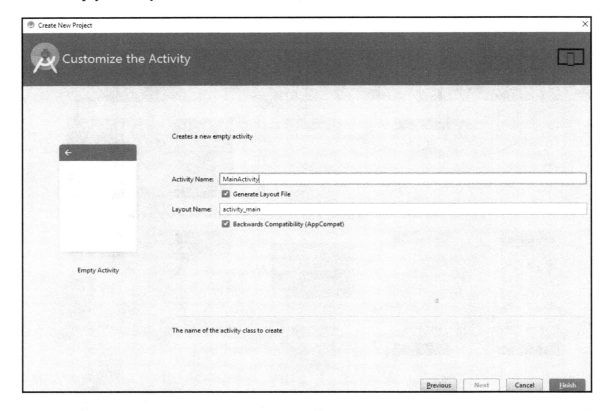

Set start-up activity name

Give a name to your activity and click on **Finish**. It may take a few minutes to set up your project. After it's done, you may see a screen like this:

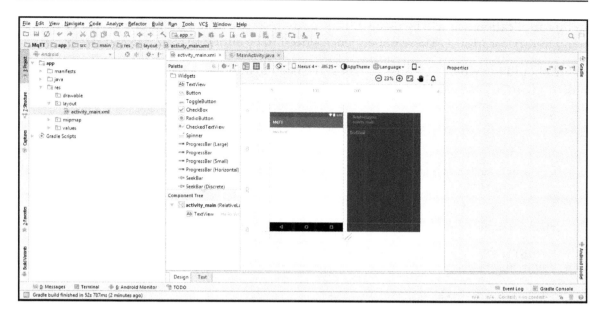

Design page. activity_name.xml

If you have a closer look over the project folder, you will notice that we have folders such as `java`, `res`, `values`, and so on. Let's have a closer look at what these folders actually contain:

- `java`: This contains all the `.java` source files for your project. The main activity, named as `MainActivity.java`, is also contained in this project.
- `res/drawable`: This is a directory for drawable components for this project. It won't be used for the moment.
- `res/layout`: This contains all the files responsible for the applications UI.
- `res/values`: This is a kind of directory for various other `xml` files that contain definitions of resources, such as string and color.
- `AndroidManifest.xaml`: This is a manifest file that defines the application as well as the permissions required by the application.
- `build.gradle`: This is an auto-generated file that contains information such as `compileSdkVersion`, `buildToolsVersion`, `applicationID`, and so on.

In this application, we will be using a third-party resource or library known as the eclipse `paho` library for MQTT. These dependencies need to be added to `build.gradle`.

There should be two `build.gradle` files. We need to add the dependencies in the `build.gradle(Module:app)` file:

```
repositories
{
  maven
    {
      url "https://repo.eclipse.org/content/repositories/paho-
      snapshots/"
    }
}
dependencies
{
  compile('org.eclipse.paho:org.eclipse.paho.android.service:1.0.3-
  SNAPSHOT')
    {
      exclude module: 'support-v4'
    }
}
```

A dependency block should already exist, so you need not write the entire thing again. In that case, just write `compile('org.eclipse.paho:org.eclipse.paho.android.service:1.0.3-SNAPSHOT') { exclude module: 'support-v4'` in the already present dependency block. Immediately after you paste the code, Android Studio will ask you to sync gradle. It is necessary that you sync gradle before proceeding:

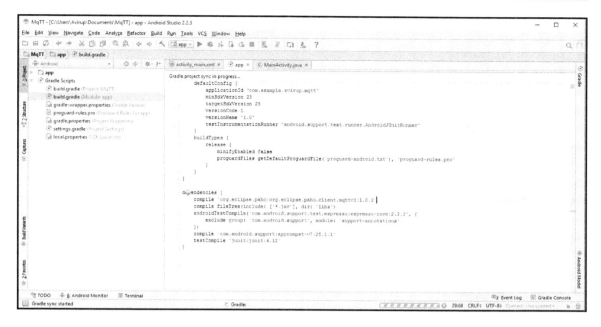

Add dependencies

Now we need to add permissions and services to our project. Browse to
`AndroidManifest.xml` and add the following permission and services:

```
<service android:name="org.eclipse.paho.android.service.MqttService" >
</service>
<uses- permission android:name="android.permission.INTERNET" />
```

After this is done, we will move forward with the UI. The UI needs to be designed under
the layout, in the `activity_main.xml` file.

We'll have the following UI components:

- `EditText`: This is for the broker
- `URL EditText`: This is for the `EditText` port for the channel

Button to connect:

- On button to send the on signal
- Off button to send the off signal

Drag and drop the previously mentioned components in the designer window. Alternatively, you can directly write it in the text view.

For your reference, the XML code of the final design is shown as follows. Write your code inside the relative layout tab:

```
<EditText
    android:layout_width="wrap_content"
    android:layout_height="wrap_content"
    android:text="android/edison"
    android:id="@+id/channelID"
    android:hint="Enter channel ID"
    android:layout_centerVertical="true"
    android:layout_alignParentStart="true"
    android:layout_alignEnd="@+id/portNum" />

<Button
    android:layout_width="wrap_content"
    android:layout_height="wrap_content"
    android:text="On"
    android:id="@+id/on"
    android:layout_below="@+id/connectMQTT"
    android:layout_alignParentStart="true"
    android:layout_marginTop="45dp" />

<Button
    android:layout_width="wrap_content"
    android:layout_height="wrap_content"
    android:text="Off"
    android:id="@+id/off"
    android:layout_alignTop="@+id/on"
    android:layout_alignParentEnd="true" />

<EditText
    android:layout_width="wrap_content"
    android:layout_height="wrap_content"
    android:id="@+id/brokerAdd"
    android:layout_alignParentTop="true"
    android:layout_alignParentStart="true"
    android:layout_marginTop="40dp"
    android:hint="Broker Address"
    android:layout_alignParentEnd="true"
    android:text="iot.eclipse.org" />

<EditText
    android:layout_width="wrap_content"
    android:layout_height="wrap_content"
    android:id="@+id/portNum"
```

```
android:layout_below="@+id/brokerAdd"
android:layout_alignParentStart="true"
android:layout_marginTop="40dp"
android:hint="Port Default: 1883"
android:layout_alignEnd="@+id/brokerAdd"
android:text="1883" />

<Button
android:layout_width="wrap_content"
android:layout_height="wrap_content"
android:text="@string/connect"
android:id="@+id/connectMQQT"
android:layout_below="@+id/channelID"
android:layout_alignParentStart="true"
android:layout_alignEnd="@+id/channelID" />
```

Now click on the **Design** view; you will see that a UI has been created, which should be somewhat similar to that of the following screenshot:

Application design

Now have a closer look at the preceding code to try to find out the properties that were used. Basic properties such as `height`, `width`, and `position` are set, which is understandable from the code. The main properties are `text`, `id` and `hint` of the `EditText`. Each component in the Android UI should have a unique ID. Beside that, we set a hint such that the user knows exactly what to enter in the text areas. For ease, we have defined the text such that while deploying, we don't need to do that again. In the final application, remove the text properties. There is another option to get your values from `strings.xml`, which can be found under values for the texts or the hints:

```
android:text="@string/connect"
```

Now that we have the UI ready, we need to implement our code that will use these UI components to interact with the device using the MQTT protocol. We also have the dependencies in place. The main Java code is written in `MainActivity.java`.

Before proceeding further with the `MainActivity.java` activity, let's create a class that will handle the MQTT connection. This will make the code a lot easier to understand and more efficient. Have a look at the following screenshot to see the location of the `MainActivity.java` file:

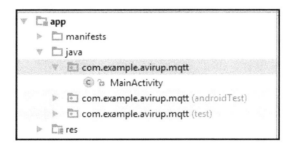

Right click on the highlighted folder and click on **new** | **java** class. This class will handle all the required data exchanges happening between the application and the MQTT broker:

```
package com.example.avirup.mqtt;
import org.eclipse.paho.client.mqttv3.IMqttDeliveryToken;
import org.eclipse.paho.client.mqttv3.MqttCallback;
import org.eclipse.paho.client.mqttv3.MqttClient;
import org.eclipse.paho.client.mqttv3.MqttException;
import org.eclipse.paho.client.mqttv3.MqttMessage;
import org.eclipse.paho.client.mqttv3.MqttPersistenceException;
import org.eclipse.paho.client.mqttv3.MqttSecurityException;
import org.eclipse.paho.client.mqttv3.persist.MemoryPersistence;
import java.io.UnsupportedEncodingException;
/** * Created by Avirup on 16-02-2017. */
public class MqttClassimplements MqttCallback
```

```
{
  String serverURI, port, clientID;
  MqttClientclient;
  MqttCallback callback;
//ConstructorMqttClass(String uri, String port, String clientID)
    {
      this.serverURI=uri;
      this.port=port;
      this.clientID=clientID;
    }
  public void MqttConnect()
    {
      try
        {
          MemoryPersistencepersistance = new MemoryPersistence();
          StringBuilderServerURI = new StringBuilder();
          ServerURI.append("tcp://");
          ServerURI.append(serverURI);
          ServerURI.append(":");
          ServerURI.append(port);
          String finalServerUri = ServerURI.toString();
          client = new MqttClient(finalServerUri, clientID,
          persistance);
          client.setCallback(callback);
          client.connect();
        }
      catch (MqttSecurityException e)
        {
          e.printStackTrace();
        }
      catch (MqttException e)
        {
          e.printStackTrace();
        }
    }
    public void MqttPublish(String message)
      {
        String commId=clientID;
        try
          {
            byte[]
            payload=message.getBytes("UTF-8");
            MqttMessagefinalMsg= new MqttMessage(payload);
            client.publish(clientID,finalMsg);
          }
        catch (UnsupportedEncodingException e)
          {
            e.printStackTrace();
```

```
        }
    catch (MqttPersistenceExceptione)
      {
        e.printStackTrace();
      }
    catch (MqttException e)
      {
        e.printStackTrace();
      }
    }
  @Override
  public void connectionLost(Throwable cause)
    {
}
    @Override
    public void messageArrived(String topic, MqttMessage
    message) throws Exception
      {
    }
  @Override
  public void deliveryComplete(IMqttDeliveryToken token)
    {
    }
  }
```

The code that is pasted earlier may seem complicated at first glance, but it's actually very simple once you understand it. It is assumed that the reader has a basic understanding of object-oriented programming concepts.

The statements that import the packages are all done automatically. After creating the class, implement the MqttCallback interface. This will add the abstract methods that are required to be overridden.

Initially, we write a parameterized constructor for this class. We also create a global reference variable for the MqttClient and the MqttCallback classes. Three global variables are also created for serverURI, port, and clientID:

```
String serverURI, port, clientID;
MqttClientclient;
MqttCallback callback;
MqttClass(String uri, String port, String clientID)
{
  this.serverURI=uri;
  this.port=port;
  this.clientID=clientID;
}
```

The parameters are the broker `URI`, `port` number, and the `clientID`.

Next, we have created three global variables that are set to the parameters. In the `MqttConnect` method, we initially form a string as we take input as just the server URI. Here, we append it with `tcp://` and the port number and also create an object for the `MemoryPersistence` class:

```
MemoryPersistencepersistance = new MemoryPersistence();
StringBuilderServerURI = new StringBuilder(); ServerURI.append("tcp://");
ServerURI.append(serverURI);
ServerURI.append(":");
ServerURI.append(port);
String finalServerUri = ServerURI.toString();
```

Next, we create the objects for the global reference variables using the new keyword:

```
client = new MqttClient(finalServerUri, clientID, persistance);
```

Please note the parameters as well.

The preceding code is surrounded by a try catch block to handle exceptions. The catch block is shown as follows:

```
catch(MqttSecurityException e)
{
  e.printStackTrace();
}
catch (MqttException e)
{
  e.printStackTrace();
}
```

The connection part is achieved. The next phase is to create the `publish` method that will publish the data to the broker.

The parameter is just the `message` of type string:

```
public void MqttPublish(String message)
{
  String commId=clientID;
  try
    {
      byte[] payload=message.getBytes("UTF-8");
      MqttMessagefinalMsg= new MqttMessage(payload);
      client.publish(clientID,finalMsg);
    }
  catch (UnsupportedEncodingException e)
```

```
      {
        e.printStackTrace();
      }
    catch (MqttPersistenceException
      {
        e.printStackTrace();
      }
    catch (MqttException e)
      {
        e.printStackTrace();
      }
}
```

`client.publish` is used to publish data. The parameter is a string which is the `clientID` or `channelID` and an object of type `MqttMessage`. `MqttMessage` contains our message. However, it doesn't accept strings. It uses a byte array. In the try block, we first convert the string to a byte array and then publish the final message by using the `MqttMessage` class to the specific channel.

For this specific application, the overridden methods aren't required, so we leave it as is.

Now head back to the `MainActivity.java` class. We will use the `MqttClass` that we just created to do the publish action. The main task here is to get data from the UI and use it to connect to the broker using the class that we just wrote.

The `MainActivity.java` will contain the following code by default:

```
packagecom.example.avirup.mqtt;
import android.support.v7.app.AppCompatActivity;
import android.os.Bundle;
public class MainActivityextends AppCompatActivity
{
  @Overrideprotected void onCreate(Bundle savedInstanceState)
  {
    super.onCreate(savedInstanceState);
    setContentView(R.layout.activity_main);
  }
}
```

Whenever the application is opened, the `onCreate` method is triggered. On having a closer look at the activity life cycle, the concept will be clear.

The life cycle callbacks are:

1. onCreate()
2. onStart()
3. onResume()
4. onPause()
5. onStop()
6. onDestroy()

More details on the life cycle can be obtained from:

```
https://developer.android.com/guide/components/activities/activity-lifecycle
.html
```

Now we need to assign some reference variables to the UI components. We'll do that on a global level.

Before the start of onCreate method, that is before the keyword override, add the following lines:

```
EditTextserverURI,port,channelID;
Button connect,on,off;
```

Now, in the onCreate method, we need to assign the reference variables we just declared and explicitly typecast them to the class type:

```
serverURI=(EditText)findViewById(R.id.brokerAdd);
port=(EditText)findViewById(R.id.port Num);
connect=(Button)findViewById(R.id.connectMQTT);
channelID=(EditText)findViewById(R.id.channelID);
on=(Button)findViewById(R.id.on);
off=( Button)findViewById(R.id.off);
```

In the preceding lines, we have explicitly type-casted them to EditText and Button, and bound them to the UI components.

Now we will create a new event handler for the connect button:

```
connect.setOnClickListener(new View.OnClickListener()
{
  @Overridepublic void onClick(View v)
    {
    }
});
```

The preceding block is activated when we press the connect button. The block contains a method whose parameter is view. The code that needs to be executed when the button is pressed needs to be written inside the onCLick (View v) method.

Before that, create a global reference variable for the class that you created before:

```
MqttClassmqttClass;
```

Next, inside the method, get the text from the edit boxes. Declare the global variables for those of the type string beforehand:

```
String serverUri, portNo,channelid;
```

Now, write the following code inside the onClick method:

```
serverUri=serverURI.getText().toString();
portNo=port.getText().toString();
channelid=channelID.getText().toString();
```

Once we get the data, we will create an object for the MqttClass class and pass the strings as parameters, and we will also invoke the MqttConnect method:

```
mqttClass=new MqttClass(serverUri,portNo,channelid);
mqttClass.MqttConnect();
```

Now we'll create similar cases for the ON and OFF methods:

```
on.setOnClickListener(new View.OnClickListener()
{
  @Overridepublic void onClick(View v)
    {
      mqttClass.MqttPublish("ON");
    }
});
off.setOnClickListener(new View.OnClickListener()
{
  @Overridepublic void onClick(View v)
    {
      mqttClass.MqttPublish("OFF");
    }
});
```

We have used the `MqttPublish` method of `MqttClass`. The parameter is just a string and is based on the `onClick` method that when it is activated, it publishes the data.

Now the application is ready and can be deployed on your device. You must turn on the developer mode on your Android device and to deploy it, connect your device to a PC and press the **Run** button. You should now have the application running on your device. To test your application, you can directly use Edison or just use the MyMqtt application.

Windows Presentation Foundation application for controlling using MQTT

WPF is a powerful UI framework for building Windows desktop client applications. It supports a broad spectrum of application features including models, controls, graphics layout, data binding, documents, and security. The programming is based on C# for the core logic and XAML for the UI.

Sample "Hello World" application in WPF

Before moving on to the development of an application for controlling Intel Edison, let's have a brief look at how we can integrate certain basic features such as a button click event, handling displaying data, and so on. Open up your Visual Studio and select **New Project**.

In PCs with low RAM, the installation of Visual Studio may take a while, as will opening Visual for the first time:

 The reason we are working with WPF is that it will be used in multiple topics, such as those in this chapter and in the upcoming chapters on robotics. In robotics, we'll be developing software to control robots. It is also assumed that the reader has an understanding of Visual Studio. For detailed information about how to work with Visual Studio and WPF, refer to the following link:

`https://msdn.microsoft.com/en-us/library/aa970268(v%3Dvs.110).aspx`

Create new project in WPF

Click on **New Project**, then under the **Visual C#** section, click on **WPF Application**. Enter a name such as `Mqtt Controller` in the field of **Name** and click on **OK**.

Once you click **OK**, the project will be created:

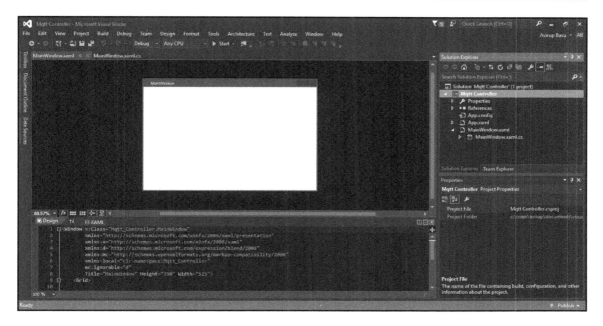

WPF project created

Once the project is created, you should get a display similar to this. If some display components are missing from your window, then go to **View** and select those. Now have a close look on the Solution Explorer, which is visible on the right-hand side of the image.

There, have a look at the project structure:

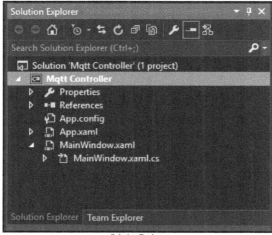

Solution Explorer

An application has two main components. The first is the UI, which will be designed in `MainWindow.xaml`, and the second is the logic, which will be implemented in `MainWindow.xaml.cs`.

The UI is designed using XAML, while the logic is implemented in C#.

To start with, we'll just have one button control: a field where the user will enter some text and an area where the entered text will be displayed. After we have a fair idea about handling events, we can move forward to the implementation of MQTT.

Initially, we'll design the UI for the double click on `MainPage.xaml.cs`. It's in this file that we'll add the UI's XAML components. The code is written in XAML and much of the work can be accomplished by the use of drag and drop feature. From the toolbox situated on the right-hand side of the application, look up the following items:

- `Button`
- `TextBlock`
- `TextBox`

There are two ways of adding the components. The first is to manually add the code in the XAML view of the page, while the second is to drag and drop from the components' toolbox. A few things to note are as follows.

The Designer window can be edited according to your wishes. A quick workaround for this is to select the component you want to edit, which can be done in the **Properties** window.

Properties can also be edited using XAML:

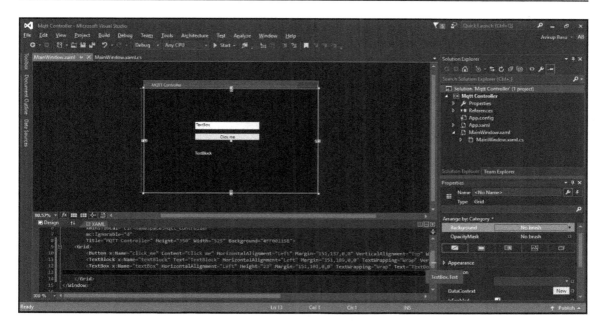

Visual Studio layout

In the preceding screenshot, we've changed the background color and added the components. Note the properties window where the background color is highlighted.

The `TextBox` is the area where the user will enter the text and the `TextBlock` is the area where it will be displayed. Once you have the components placed on the design view and have edited their properties, mainly the names of the components, we'll add the event handlers. For a shortcut of the design shown in the preceding screenshot, write the following XAML code within the `grid` tag:

```
<Button x:Name="click_me" Content="Click me" HorizontalAlignment="Left"
Margin="151,137,0,0" VerticalAlignment="Top" Width="193"/>
<TextBlock x:Name="textBlock" Text="TextBlock" HorizontalAlignment="Left"
Margin="151,189,0,0" TextWrapping="Wrap" VerticalAlignment="Top"
Width="193" Foreground="White"/>
<TextBox x:Name="textBox" HorizontalAlignment="Left" Height="23"
Margin="151,101,0,0" TextWrapping="Wrap" Text="TextBox"
VerticalAlignment="Top" Width="193"/>
```

Now in the **Designer** window, double click on the button to create an event handler for a click event. The events that are available can be viewed in the **Properties** window, as shown in the following screenshot:

Event properties of button

Once you have double clicked, you will automatically be redirected to `MainWindow.xaml.cs` along with a method self-generated for the event.

You will get a method something similar to the following code:

```
privatevoidclick_me_Click(object sender, RoutedEventArgs e)
{
}
```

Here, we are going to implement the logic. Initially, we will read the data as written in the `TextBox`. If it's empty, we'll display a message saying that it cannot be empty. Then, we'll just pass the message to the `TextBlock`. The following code does the same thing:

```
privatevoidclick_me_Click(object sender, RoutedEventArgs e)
{
   string res = textBox.Text; if(string.IsNullOrEmpty(res))
      {
         MessageBox.Show("No text entered. Please enter again");
      }
   else
      {
         textBlock.Text = res;
      }
}
```

The preceding code initially reads the data and then checks if it's null or empty and then outputs the data into the `TextBlock`:

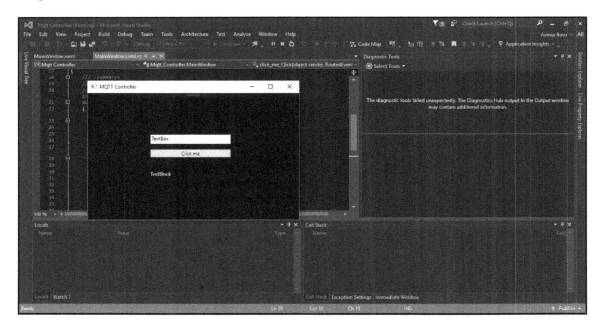

Application run—1

Press *F5* to run your application and then the preceding screens should appear. Next, delete the text in the **TextBox** and click on the **Click me** button:

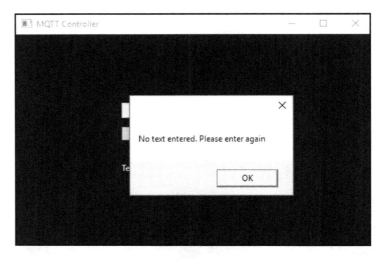

Empty text

Now, enter any text in the **TextBox** and press the **Click me** button. Your entered text should be displayed following in the **TextBlock**:

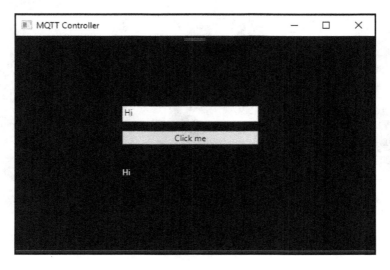

WPF HelloWorld

Now that we know how to make a simple WPF application, we are going to edit the application itself to implement the MQTT protocol. To implement the MQTT protocol, we have to use a library, which will be added using the nugget package manager.

Now browse to **References** and click on **Manage Nugget Packages** and add the M2Mqtt external library:

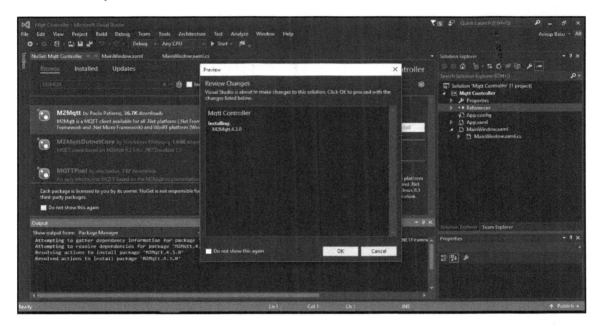

NuGet package manager

Once we have the packages, we can use them in our project. For this project, we'll be using the following UI components in `MainWindow.xaml`:

- A **TextBox** for entering the channel ID
- A **TextBlock** to display the latest control command
- A button to set the status as on
- A button to set the status as **off**
- A button to **Connect**

Feel free to design the UI on your own:

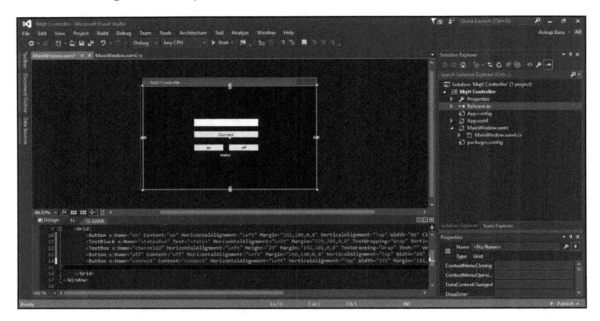

UI for controller application

In the preceding screenshot, you will see that the design is updated and a button has also been added. The code for the preceding design is pasted as follows. The **TextBox** is the area where we'll enter the channel ID, and then we will use the buttons to turn an LED **on** and **off**, and the **Connect** button to connect to the service. Now, as done previously, we will create event handlers for click events for the two buttons mentioned previously. To add click events, simply double click on each button:

```
<Button x:Name="on" Content="on" HorizontalAlignment="Left"
Margin="151,180,0,0" VerticalAlignment="Top" Width="88" Click="on_Click"/>
   <TextBlock x:Name="statusBox" Text="status"
   HorizontalAlignment="Left" Margin="229,205,0,0" TextWrapping="Wrap"
   VerticalAlignment="Top" Width="115" Foreground="White"/>
   <TextBox x:Name="channelID" HorizontalAlignment="Left" Height="23"
   Margin="151,101,0,0" TextWrapping="Wrap" Text=""
   VerticalAlignment="Top" Width="193"/>
   <Button x:Name="off" Content="off" HorizontalAlignment="Left"
   Margin="256,180,0,0" VerticalAlignment="Top" Width="88"
   Click="off_Click"/>
   <Button x:Name="connect" Content="Connect" HorizontalAlignment="Left"
   VerticalAlignment="Top" Width="193" Margin="151,139,0,0"
   Click="connect_Click"/>
```

The preceding code is mentioned in the grid tag.

Now, once you have the design, move on to `MainWindow.xaml.cs` and write the main code. You will notice that a constructor and two event handler methods already exist.

Add the following namespace to use the library using:

```
uPLibrary.Networking.M2Mqtt;
```

Now create an instance of the `MqttClient` class and declare a global string variable:

```
MqttClient client = new MqttClient("iot.eclipse.org");
String channelID;
```

Next, in the event handler for the **Connect** button, connect it to the broker using the channel ID.

The entire code for the **Connect** button's event handler is mentioned as follows:

```
channelID_text = channelID.Text;
if (string.IsNullOrEmpty(channelID_text))
{
  MessageBox.Show("Channel ID cannot be null");
}
else
{
  try
  {
    client.Connect(channelID_text); connect.Content = "Connected";
  }
catch (Exception ex)
{
  MessageBox.Show("Some issues occured: " + ex.ToString());
}
}
```

In the preceding snippet, we read the data from the `textbox` that contains the channel ID. If it's null, we ask the user to enter it again. Then, finally, we connect it to the channel ID. Note that it is inside the `try catch` block.

There are two more event handlers. We need to publish some value to the channel they are connected to.

In the on button's event handler, insert the following code:

```
private void on_Click(object sender, RoutedEventArgs e)
{
  byte[] array = Encoding.ASCII.GetBytes("on");
  client.Publish(channelID_text, array);
}
```

As seen in the preceding code, the parameter for the Publish method is the topic, which is the channelID and a byte[] array which contains the message.

Similarly, for the off method, we have:

```
private void off_Click(object sender, RoutedEventArgs e)
{
  byte[] array = Encoding.ASCII.GetBytes("off");
  client.Publish(channelID_text, array);
}
```

That's it. That's the entire code for your MQTT controller for home automation. The entire code is pasted as follows for your reference:

```
using System;
usingSystem.Collections.Generic;
usingSystem.Linq;
usingSystem.Text;
usingSystem.Threading.Tasks;
usingSystem.Windows;
usingSystem.Windows.Controls;
usingSystem.Windows.Data;
usingSystem.Windows.Documents;
usingSystem.Windows.Input;
usingSystem.Windows.Media;
usingSystem.Windows.Media.Imaging;
usingSystem.Windows.Navigation;
usingSystem.Windows.Shapes;
using uPLibrary.Networking.M2Mqtt;
namespaceMqtt_Controller
{
/// <summary>
/// Interaction logic for MainWindow.xaml
/// </summary>
  public partial class MainWindow : Window
    {
      MqttClient client = new MqttClient("iot.eclipse.org");
      String channelID_text;
      publicMainWindow()
        {
```

```
              InitializeComponent();
       }
    private void on_Click(object sender, RoutedEventArgs e)
       {
          byte[] array = Encoding.ASCII.GetBytes("ON");
          client.Publish(channelID_text, array);
          statusBox.Text = "on";
       }
    private void off_Click(object sender, RoutedEventArgs e)
       {
          byte[] array = Encoding.ASCII.GetBytes("OFF");
          client.Publish(channelID_text, array);
          statusBox.Text = "off";
       }
    private void connect_Click(object sender, RoutedEventArgs e)
       {
          channelID_text = channelID.Text;
          if (string.IsNullOrEmpty(channelID_text))
             {
               MessageBox.Show("Channel ID cannot be null");
             }
          else
             {
               try
                  {
                     client.Connect(channelID_text);
                     connect.Content = "Connected";
                  }
               catch (Exception ex)
                  {
                     MessageBox.Show("Some issues occured: " +
                     ex.ToString());
                  }
             }
       }
    }
}
}
```

Press the *F5* or the **Start** button to execute this code:

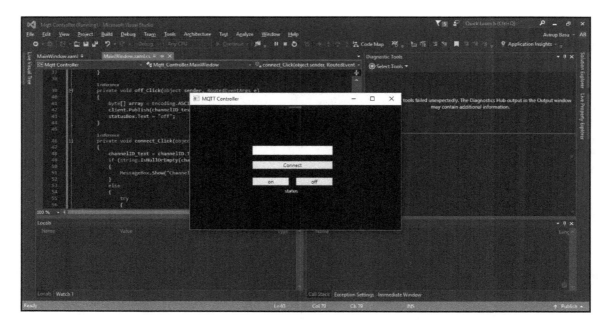

Application running

Next, in the **TextBox**, enter the `channelID`. Here, we'll be entering it as `avirup/control` and then we will press the **Connect** button:

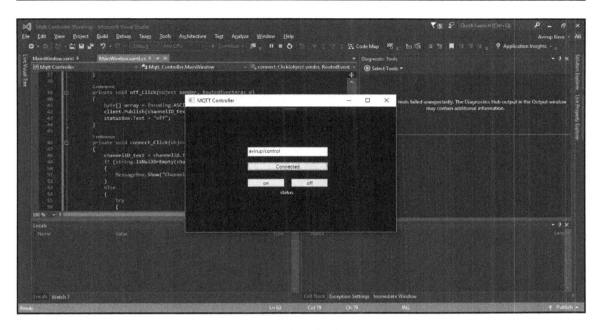

Application running—2

Now open your PuTTY console and log in to Intel Edison. Verify that the device is connected to the Internet using the `ifconfig` command. Next, just run the Node.js script. Next, press the ON button:

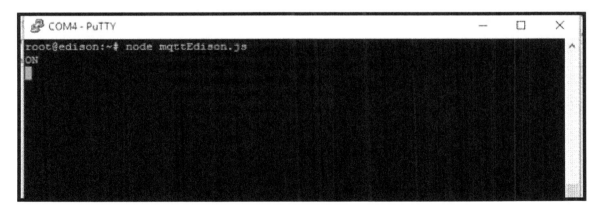

MQTT controlled by WPF application

Similarly, on pressing the OFF button, you will see a screen similar to the following:

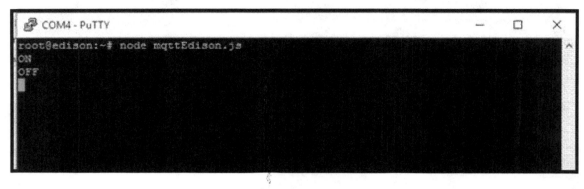

MQTT controlled by WPF

Keep pressing ON and OFF and you will see the effect on Intel Edison. Now that we remember that we have connected the relay and the electric bulb, the effect should be visible by now. If the main switch of the AC power supply is turned off, then you won't see the bulb getting turned on, but you will hear a `tick` sound. That suggest that the relay is now in the ON position. The image of the hardware setup is shown as follows:

Hardware setup for home automation

Thus, you have a home automation setup ready and you can control it by the PC application or the Android application.

 If you are in office network, then sometimes port 1883 is blocked. In those cases, it is recommended to use your own personal network.

Open-ended task for the reader

Now, you may have got a brief idea about how things must work in home automation. We have covered multiple areas in this niche. The task that is left for the reader is not only to integrate a single control command, but multiple control commands. This will allow you to control multiple devices. Add more functionality in the Android and the WPF application and go with more string control commands. Connect more relay units to the device for interfacing.

Summary

In this chapter, we've learned about the idea of home automation in its crude form. We also learned about how we can control an electrical load using relays. Not only that, but also we learned how to develop a WPF application and implement the MQTT protocol. On the device end, we used a Node.js code to connect our device to the Internet and subscribe to certain channels using the broker and ultimately receive signals to control itself. In the Android side of the system, we have used an already available MyMqtt application and used it to both to get and publish data. However, we also covered the development of the Android application in detail and showcased the use of it in implementing the MQTT protocol to control devices.

In Chapter 4, *Intel Edison and Security System*, we are going to learn how to deal with image processing and speech processing applications using Intel Edison. Chapter 4, *Intel Edison and Security System*, will mainly deal with Python and the usage of some open source libraries.

4
Intel Edison and Security System

In previous chapters, we learned how we can use the Intel Edison to develop applications related to IoT where we displayed live sensor data and also controlled the Edison itself. We also learned the development of an Android and a WPF app that was used to control the Intel Edison. Well, this chapter is more on the local front of the Intel Edison where we are going to use the built-in features of the device. This chapter is concentrated mainly on two key points:

- Speech and voice processing with the Intel Edison
- Image processing with the Intel Edison

All the codes are to be written in Python, so some parts of the chapter will concentrate on Python programming as well. In this chapter, we'll operate the Intel Edison using voice commands and then ultimately detect faces using the Intel Edison and a webcam. This chapter will thus explore the core capabilities of the Intel Edison. Since most of the code is in Python, it is advisable to download Python for your PC from the following website:

```
https://www.python.org/downloads/
```

This chapter will be divided into two parts. The first part will concentrate on only speech or voice processing and we'll do a mini-project based on that while the second part, which will be a bit lengthy, will concentrate on the image processing aspect of it using OpenCV.

Speech/voice processing using Edison

Speech processing typically refers to the various mathematical techniques that are applied on an audio signal to process it. It may be some simple mathematical operation or some complex operation. It's a special case of digital signal processing. However, we are not typically dealing with speech processing as a whole entity. We are interested only in a specific area of speech to text conversion. It is to be noted that everything in this chapter is to be performed by the Edison itself without accessing any cloud services. The scenario that this chapter will tackle initially is that we'll make the Edison perform some tasks based on our voice commands. We'll be using a lightweight speech processing tool, but before we proceed further with all the code and circuits, make sure you have the following devices with you. Initially, we'll walk you through switching an LED on and off. Next we'll deal with controlling a servo motor using voice commands.

Devices required

Along with the Intel Edison, we need a couple of more devices, as listed here:

- Power adapter of 9V-1 A for the Intel Edison
- USB sound card
- USB hub, preferably powered

This project will use the Edison on external power and the USB port will be used for the sound card. A non-powered USB hub also works, but because of the current it's recommended to use a powered USB hub.

 Make sure that the USB sound card is supported on a Linux environment. The selector switch should be towards the USB port. That is because the Edison will be powered through the DC adapter and we need power in the USB port that is activated only when we provide DC power.

Speech processing library

For this project we are going to use PocketSphinx. It's a lightweight version of CMU Sphinx, a project created by Carnegie Mellon University. It's a lightweight speech recognition engine meant for mobile and handheld devices and wearables. The greatest advantage of using this over any cloud-based service is that it is available offline.

More information about PocketSphinx can be accessed from the following links:

```
http://cmusphinx.sourceforge.net/wiki/develop
```

```
https://github.com/cmusphinx/pocketsphinx
```

Setting up the library will be discussed in a later section of this chapter.

Initial configuration

In the first chapter, we performed some very basic configuration for the Intel Edison. Here we need to configure our device with the required libraries and sound setup with the sound card. For this you need to connect the Intel Edison to only one micro USB port. This will be used to communicate using the PuTTY console and transfer files using the FileZilla FTP client:

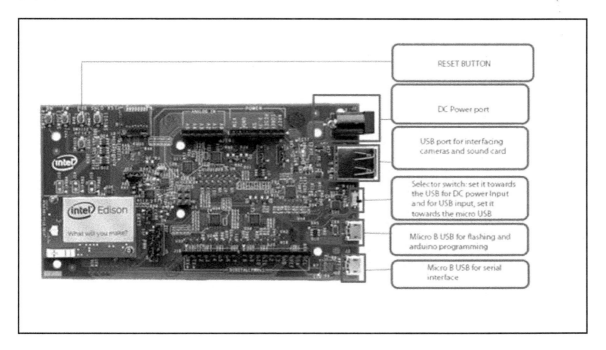

Arduino expansion board components

Connect the Intel Edison to the Micro B USB for serial interface to your PC.

Some of the steps were covered in `Chapter 1`, *Setting up Intel Edison;* however, we'll show all the steps from the beginning. Open your PuTTY console and log in to your device. Use the `configure_edison -wifi` to connect to your Wi-Fi network.

Initially, we'll add AlexT's unofficial `opkg` repository. To add this, edit the `/etc/opkg/base-feeds.conf` file.

Add these lines to the preceding file:

```
src/gz all http://repo.opkg.net/edison/repo/all
src/gz edison http://repo.opkg.net/edison/repo/edison
src/gz core2-32 http://repo.opkg.net/edison/repo/core2-32
```

To do that, execute the following command:

```
echo "src/gz all http://repo.opkg.net/edison/repo/all
src/gz edison http://repo.opkg.net/edison/repo/edison
src/gz core2-32 http://repo.opkg.net/edison/repo/core2-32" >>
/etc/opkg/base-feeds.conf
```

Update the package manager:

```
opkg update
```

Install `git` using the package manager:

```
opkg install git
```

We will now install Edison helper scripts to simplify things a bit:

1. First `clone` the package:

   ```
   git clone https://github.com/drejkim/edison-scripts.git ~/edison
   scripts
   ```

2. Now we have to add `~/edison-scripts` to the path:

   ```
   echo'export PATH=$PATH:~/edison-scripts'>>~/.profile
   source~/.profile
   ```

3. Next we will run the following scripts:

```
# Resize /boot -- we need the extra space to add an additional
kernel

    resizeBoot.sh


# Install pip, Python's package manager

    installPip.sh


# Install MRAA, the low level skeleton library for IO
communication on, Edison, and other platforms

    installMraa.sh
```

The initial configuration is done. Now we'll configure the Edison for sound.

4. Now `install` the modules for USB devices, including USB webcams, microphone, and speakers. Make sure that your sound card is connected to the Intel Edison:

```
opkg install kernel-modules
```

5. The next target is to check whether the USB device is getting detected or not. To check that, type the `lsusb` command:

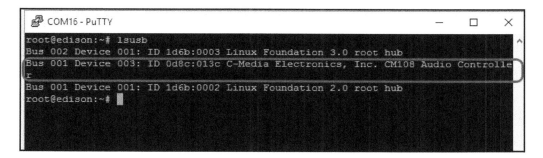

USB sound card

The device that is connected to the Intel Edison is shown in the preceding screenshot. It is highlighted in the box. Once we get the device that is connected to the Edison, we can proceed further.

Now we'll check whether `alsa` is able to detect the sound card or not. Type in the following command:

```
aplay -Ll
```

```
COM16 - PuTTY                                                      —    □    ✕
  Subdevice #1: subdevice #1
  Subdevice #2: subdevice #2
  Subdevice #3: subdevice #3
  Subdevice #4: subdevice #4
  Subdevice #5: subdevice #5
  Subdevice #6: subdevice #6
  Subdevice #7: subdevice #7
card 0: Loopback [Loopback], device 1: Loopback PCM [Loopback PCM]
  Subdevices: 8/8
  Subdevice #0: subdevice #0
  Subdevice #1: subdevice #1
  Subdevice #2: subdevice #2
  Subdevice #3: subdevice #3
  Subdevice #4: subdevice #4
  Subdevice #5: subdevice #5
  Subdevice #6: subdevice #6
  Subdevice #7: subdevice #7
card 1: dummyaudio [dummy-audio], device 0: 14 []
  Subdevices: 1/1
  Subdevice #0: subdevice #0
card 2: Device [USB PnP Sound Device], device 0: USB Audio [USB Audio]
  Subdevices: 1/1
  Subdevice #0: subdevice #0
root@edison:~#
```

Alsa device check

It is noted that our device is getting detected as card 2, named as `Device`.

Now we have to create a `~/.asoundrc` file where we need to add the following line. Please note that `Device` must be replaced with the device name that is detected on your system:

```
pcm.!default sysdefault:Device
```

Now, once this is done, exit and save the file. Next, to test whether everything is working or not, execute the following command and you must hear something on the headphone connected:

```
aplay /usr/share/sounds/alsa/Front_Center.wav
```

You should hear the words `Front Center`.

Now, our target is to record something and interpret the result. So let's test whether recording is working or not.

To record a clip, type in the following command:

```
arecord ~/test.wav
```

Press *Ctrl* + *c* to stop recording. To play the preceding recording, type the following:

```
aplay ~/test.wav
```

You must hear what you have recorded. If you are not able to hear the sound, type `alsamixer` and adjust the playback and record volumes. Initially, you need to select the device:

```
lqqqqqqqqqqqqqqqqqqqqqqqqqqqqq AlsaMixer v1.0.28 qqqqqqqqqqqqqqqqqqqqqqqqqqqqqk
x Card:  Loopback                                 F1:   Help                  x
x Chip:  Loopback Mixer                           F2:   System information    x
x View:  F3:[Playback] F4: Capture  F5: All       F6:   Select sound card     x
x Item:  PCM [dB gain: 0.00, 0.00]                Esc:  Exit                  x
x                                                                             x
x                                                                             x
x                             lqqk                                            x
x                             xaax                                            x
x                  lqqqqq Sound Card qqqqqqk                                  x
x                  x-   (default)         x                                   x
x                  x0   Loopback          x                                   x
x                  x1   dummy-audio       x                                   x
x                  x2   USB PnP Sound Devicex                                 x
x                  x    enter device name...x                                 x
x                  mqqqqqqqqqqqqqqqqqqqqqqqqj                                  x
x                             xaax                                            x
x                             xaax                                            x
x                             xaax                                            x
x                             xaax                                            x
x                             mqqj                                            x
x                          100<>100                                          x
x                          <  PCM   >                                         x
x                                                                             x
mqqqqqqqqqqqqqqqqqqqqqqqqqqqqqqqqqqqqqqqqqqqqqqqqqqqqqqqqqqqqqqqqqqqqqqqqqqqqqqqj
```

Alsamixer—1

Next, adjust the volume using the arrow keys:

Alsamixer—2

Now everything related to sound is set up. The next aim is to `install` the packages for speech recognition.

Initially, use Python's `pip` to `install cython`:

```
pip install cython
```

The preceding package takes a lot of time to install. Once that's done, there are some shell scripts that are required to be executed. I have created a GitHub repository for this that contains the required files and the code. Use the git command to clone the repository (`https://github.com/avirup171/Voice-Recognition-using-Intel-Edison.git`):

```
git clone
```

Next in the bin folder, you will find the packages. Before typing the commands to execute those shell scripts, we need to provide permissions. Type the following command to add permissions:

```
chmod +x <FILE_NAME>
```

Next type the filename to execute them. Installation of the packages may take a bit of time:

```
./installSphinxbase.sh
```

Next type these for adding to the path:

```
echo 'export LD_LIBRARY_PATH=/usr/local/lib' >> ~/.profile
echo 'export PKG_CONFIG_PATH=/usr/local/lib/pkgconfig' >> ~/.profile
source ~/.profile
```

Next install Pocketsphinx:

```
./installPocketsphinx.sh
```

Finally, install PyAudio:

```
./installPyAudio.sh
```

After this step, all the configurations are set up and we are good to go with the coding. PocketSphinx works with some specific sets of commands. We need to create a language mode and a dictionary for the words to be used. We'll do that using the Sphinx knowledge base tool:

http://www.speech.cs.cmu.edu/tools/lmtool-new.html

Upload a text file containing the set of commands that we want the engine to decode. Then click on **COMPILE KNOWLEDGE BASE**. Download the .tgz file that contains the necessary files that are required. Once we have those files, copy it to the Edison using FileZilla. Note the names of the files that contain the following extension. Ideally each file should have the same name:

- .dic
- .lm

Move the entire set to the Edison.

Writing the code

Problem statement: To turn on and off an LED using voice commands such as ON and OFF.

Before writing the code, let us discuss the algorithm first. Please note that I am writing the algorithm in plain text so that it is easier for the reader to understand.

Let's start with the algorithm

Perform the following steps to begin with the algorithm:

1. Import all the necessary packages.
2. Set the LED pin.
3. Start an infinite loop. From here on, all the parts or blocks will be inside the while loop.
4. Store two variables in the path for the .lm and .dic files.
5. Record and save a .wav file for 3 seconds.
6. Pass the .wav file as a parameter to the speech recognition engine.
7. Get the resultant text.
8. With an if else block test for the ON and OFF texts and use the mraa library to turn on and off an LED.

The algorithm is pretty much straightforward. Compare the following code with the preceding algorithm to get a full grip of it:

```
import collections
import mraa
import os
import sys
import time

# Import things for pocketsphinx
import pyaudio
import wave
import pocketsphinx as ps
import sphinxbase

led = mraa.Gpio(13)
led.dir(mraa.DIR_OUT)


print("Starting")
```

```
while 1:
        #PocketSphinx parameters
        LMD   = "/home/root/vcreg/5608.lm"
        DICTD = "/home/root/vcreg/5608.dic"
        CHUNK = 1024
        FORMAT = pyaudio.paInt16
        CHANNELS = 1
        RATE = 16000
        RECORD_SECONDS = 3
        PATH = 'vcreg'
        p = pyaudio.PyAudio()
        speech_rec = ps.Decoder(lm=LMD, dict=DICTD)
        #Record audio
        stream = p.open(format=FORMAT, channels=CHANNELS, rate=RATE,
        input=True, frames_per_buffer=CHUNK)
        print("* recording")
        frames = []
        fori in range(0, int(RATE / CHUNK * RECORD_SECONDS)):
                data = stream.read(CHUNK)
                frames.append(data)
        print("* done recording")
        stream.stop_stream()
        stream.close()
        p.terminate()
        # Write .wav file
        fn = "test.wav"
        #wf = wave.open(os.path.join(PATH, fn), 'wb')
        wf = wave.open(fn, 'wb')
        wf.setnchannels(CHANNELS)
        wf.setsampwidth(p.get_sample_size(FORMAT))
        wf.setframerate(RATE)
        wf.writeframes(b''.join(frames))
        wf.close()

        # Decode speech
        #wav_file = os.path.join(PATH, fn)
        wav_file=fn
        wav_file = file(wav_file,'rb')
        wav_file.seek(44)
        speech_rec.decode_raw(wav_file)
        result = speech_rec.get_hyp()
        recognised= result[0]
        print("* LED section begins")
        print(recognised)
        if recognised == 'ON.':
                led.write(1)
        else:
                led.write(0)
```

```
cm = 'espeak "'+recognised+'"'
os.system(cm)
```

Let's go line by line:

```
import collections
import mraa
import os
import sys
import time

# Import things for pocketsphinx
import pyaudio
import wave
import pocketsphinx as ps
import Sphinxbase
```

The preceding segment is just to `import` all the libraries and packages:

```
led = mraa.Gpio(13)
led.dir(mraa.DIR_OUT)
```

We set the LED pin and set its direction as the output. Next we will begin the infinite while loop:

```
#PocketSphinx and Audio recording parameters
        LMD   = "/home/root/vcreg/5608.lm"
        DICTD = "/home/root/vcreg/5608.dic"
        CHUNK = 1024
        FORMAT = pyaudio.paInt16
        CHANNELS = 1
        RATE = 16000
        RECORD_SECONDS = 3
        PATH = 'vcreg'
        p = pyaudio.PyAudio()
        speech_rec = ps.Decoder(lm=LMD, dict=DICTD)
```

The preceding chunk of code is just the parameters for PocketSphinx and for audio recording. We will be recording for 3 seconds. We have also provided the path for the `.lmd` and `.dic` files and some other audio recording parameters:

```
#Record audio
        stream = p.open(format=FORMAT, channels=CHANNELS, rate=RATE,
        input=True, frames_per_buffer=CHUNK)
        print("* recording")
        frames = []
        fori in range(0, int(RATE / CHUNK * RECORD_SECONDS)):
                data = stream.read(CHUNK)
```

```
        frames.append(data)
    print("* done recording")
    stream.stop_stream()
    stream.close()
    p.terminate()
```

In the preceding code, we record the audio for the specific time interval.

Next, we save it as a .wav file:

```
# Write .wav file
    fn = "test.wav"
    #wf = wave.open(os.path.join(PATH, fn), 'wb')
    wf = wave.open(fn, 'wb')
    wf.setnchannels(CHANNELS)
    wf.setsampwidth(p.get_sample_size(FORMAT))
    wf.setframerate(RATE)
    wf.writeframes(b''.join(frames))
    wf.close()
```

The final step contains the decoding of the file and comparing it to affect the LED:

```
# Decode speech
    #wav_file = os.path.join(PATH, fn)
    wav_file=fn
    wav_file = file(wav_file,'rb')
    wav_file.seek(44)
    speech_rec.decode_raw(wav_file)
    result = speech_rec.get_hyp()
    recognised= result[0]
    print("* LED section begins")
    print(recognised)
    if recognised == 'ON.':
        led.write(1)
    else:
        led.write(0)
    cm = 'espeak "'+recognised+'"'
    os.system(cm)
```

In the preceding code, we initially pass the .wav file as a parameter to the speech processing engine and then use the result to compare the output. Finally, we switch on and off the LEDs based on the output of the speech processing engine. Another activity carried out by the preceding code is that whatever is recognized is spoken back using espeak. espeak is a text to speech engine. It uses spectral formant synthesis by default, which sounds robotic, but can be configured to use Klatt formant synthesis or MBROLA to give it a more natural sound.

Transfer the code to your device using FileZilla. Let's assume that the code is saved by the file named `VoiceRecognitionTest.py`.

Before executing the code, you may want to attach an LED to GPIO pin 13 or just use the on board LED for the purpose.

To execute the code, type the following:

```
python VoiceRecognitionTest.py
```

Initially, the console says `*recording`, speak on:

Voice recognition—1

Then, after you speak, the speech recognition engine will recognize the word that you spoke from the existing language model:

Voice recognition—2

It is noted that on is displayed. That means that the speech recognition engine has successfully decoded the speech we just spoke. Similarly, the other option stands when we speak off on the microphone:

Voice recognition—3

So now we have a voice recognition proof of concept ready. Now, we are going to use this concept with small modifications to lock and unlock the door.

Door lock/unlock based on voice commands

In this section, we'll just open and close a door based on voice commands. Similar to the previous section, where we switched an LED on and off using voice commands such as ON and OFF, here we are going to do a similar thing using a servo motor. The main target is to make the readers understand the core concepts of the Intel Edison where we use voice commands to perform different tasks. The question may arise, why are we using servo motors?

A servo motor, unlike normal DC motors, rotates up to a specific angle set by the operator. In normal scenarios, controlling the lock of a door may use a relay. The usage of relays was discussed in Chapter 3, *Intel Edison and IoT (Home Automation)*.

Let us also explore the use of servo motors so that we can widen the spectrum of controlling devices. In this case, when a servo is set to 0 degrees, it is unlocked and when it is set to 90 degrees, it is locked. The control of servo motors requires the use of pulse width modulation pins. Intel Edison has four PWM pins:

Servo motor. Picture credits: https://circuitdigest.com

There are three operating lines to operate a servo:

- Vcc
- Gnd
- Signal

The typical color coding goes like this:

- Black—ground
- Red or brown—power supply
- Yellow or white—control signal

We are using a 5V servo motor; therefore the Edison is enough to supply power. The Edison and the servo motor must share a common ground. Finally, the signal pin is connected to the PWM pin on the Intel Edison. As we move further with this mini-project, things will get clearer.

Circuit diagram

The following is the circuit diagram for voice recognition:

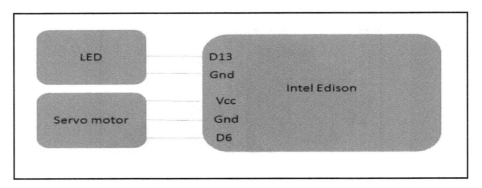

Circuit diagram for voice recognition

As already mentioned, the servo motor requires PWM pins for operation and the Intel Edison has a total of six PWM pins. Here we are using digital pin 6 for servo control and digital pin 13 for the LED. As far as the peripheral devices are concerned, connect your USB sound card to the USB of the Intel Edison and you are all set.

Configuring the servo library for Python

To control a servo, we need to send some signals through the PWM pins. We opt for using a library for controlling the servo motors.

Use the following link to get access to a `Servo.py` Python script from a GitHub repository:

`https://github.com/MakersTeam/Edison/blob/master/Python-Examples/Servo/Servo.py`

Download the file and push it to your Edison device. After that, just execute the file similar to executing a Python script:

```
python Servo.py
```

Now once you have done, you are ready to use the servo with your Intel Edison using Python.

Getting back to the hardware, the servo must be connected to digital pin 6, which is a PWM pin. Let's write a Python script that will test if the library is functioning or not:

```python
from Servo import *
import time
myServo = Servo("Servo")
myServo.attach(6)
while True:
    # From 0 to 180 degrees
    for angle in range(0,180):
        myServo.write(angle)
        time.sleep(0.005)
    # From 180 to 0 degrees
    for angle in range(180,-1,-1):
        myServo.write(angle)
        time.sleep(0.005)
```

The preceding code basically sweeps from 0 to 180 degrees and back to 0 degrees. The circuit remains the same as discussed before. Initially, we attach the servo to the servo pin. Then as the standard goes, we put the entire logic in an infinite loop. To rotate the servo to a specific angle, we use .write(angle). In the two for loops, initially we rotate from 0 to 180 degrees and in the second one, we rotate from 180 to 0 degrees.

It is also to be noted that time.sleep(time_interval) is used to pause the code for some miliseconds. When you execute the preceding code, the servo should rotate and come back to the initial position.

Now, we have all the things in place. We'll just put them in the right place and your voice controlled door will be ready. Initially, we controlled an LED and then we learned how we can operate a servo using Python. Now let's create a language model using the Sphinx knowledge base tool.

Language model

For this project, we'll be using the following set of commands. To keep things simple, we're using only two sets of commands:

- door open
- door close

Follow the process that was discussed earlier and create a text file and just write the three unique words:

door open close

Save it and upload it to the Sphinx knowledge base tool and compile it.

Once you have the compressed file downloaded, move on to the next step with this code:

```
import collections
import mraa
import os
import sys
import time

# Import things for pocketsphinx
import pyaudio
import wave
import pocketsphinx as ps
import sphinxbase
# Import for Servo
from Servo import *

led = mraa.Gpio(13)
led.dir(mraa.DIR_OUT)
myServo = Servo("First Servo")
myServo.attach(6)

print("Starting")
while 1:
        #PocketSphinx parameters
        LMD   = "/home/root/Voice-Recognition-using-Intel-Edison/8578.lm"
        DICTD = "/home/root/Voice-Recognition-using-Intel-Edison/8578.dic"
        CHUNK = 1024
        FORMAT = pyaudio.paInt16
        CHANNELS = 1
        RATE = 16000
        RECORD_SECONDS = 3
        PATH = 'Voice-Recognition-using-Intel-Edison'
        p = pyaudio.PyAudio()
        speech_rec = ps.Decoder(lm=LMD, dict=DICTD)
        #Record audio
        stream = p.open(format=FORMAT, channels=CHANNELS, rate=RATE,
input=True, frames_per_buffer=CHUNK)
        print("* recording")
        frames = []
        fori in range(0, int(RATE / CHUNK * RECORD_SECONDS)):
            data = stream.read(CHUNK)
```

```
            frames.append(data)
print("* done recording")
stream.stop_stream()
stream.close()
p.terminate()
# Write .wav file
fn = "test.wav"
#wf = wave.open(os.path.join(PATH, fn), 'wb')
wf = wave.open(fn, 'wb')
wf.setnchannels(CHANNELS)
wf.setsampwidth(p.get_sample_size(FORMAT))
wf.setframerate(RATE)
wf.writeframes(b''.join(frames))
wf.close()

# Decode speech
#wav_file = os.path.join(PATH, fn)
wav_file=fn
wav_file = file(wav_file,'rb')
wav_file.seek(44)
speech_rec.decode_raw(wav_file)
result = speech_rec.get_hyp()
recognised= result[0]
print("* LED section begins")
print(recognised)
ifrecognised == 'DOOR OPEN':
        led.write(1)
        myServo.write(90)
else:
        led.write(0)
        myServo.write(0)
cm = 'espeak "'+recognised+'"'
os.system(cm)
```

The preceding code is more or less similar to the code for switching an LED on and off. The only difference is that the servo control mechanism is added into the existing code. In a simple if else block, we check for the door open and door close conditions. Finally based on what is triggered, we set the LED and the servo to a 90 degrees or 0 degree position.

Conclusion of speech processing using the Intel Edison

From the projects discussed before, we explored one of the core capabilities of the Intel Edison and explored a whole new scenario of controlling the Intel Edison by voice. A popular use case that implements the preceding procedure can be the case of home automation, which was implemented in the earlier chapter. Another use case is building a virtual voice based assistant using your Intel Edison. There are multiple opportunities that can be used using voice-based control. It's up to the reader's imagination as to what they want to explore.

In the next part, we'll be dealing with the implementation of image processing using the Intel Edison.

Image processing using the Intel Edison

Image processing or computer vision is one such field that requires tremendous amounts of research. However, we're not going to do rocket science here. We are opting for an open source computer vision library called OpenCV. OpenCV supports multiple languages and we are going to use Python as our programming language to perform face detection.

Typically, an image processing application has an input image; we process the input image and we get an output processed image.

Intel Edison doesn't have a display unit. So essentially we will run the Python script on our PC first. Then after the successful working of the code in the PC, we'll modify the code to run on the Edison. Things will get clearer when we do the practical implementation.

Our target is to perform face detection and, if detected, perform some action.

Initial configuration

The initial configuration will include installing the openCV package both on the Edison device as well as the PC.

For the PC, download Python from `https://www.python.org/downloads/windows/`. Next install Python on your system. Also download the latest version of openCV from `https://sourceforge.net/projects/opencvlibrary/`.

After you download openCV, move the extracted folder to C:\. Next, browse to C:\opencv\build\python\2.7\x86.

Finally, copy the cv2.pyd file to C:\Python27\Lib\site-packages.

We need to install numpy as well. Numpy stands for **Numerical Python**. Download and install it.

Once you install all the components, we need to test whether everything is installed or not. To do that, open up the idle Python GUI and type the following:

```
importnumpy
import cv2
```

If this proceeds without any error, then everything is installed and in place as far as the PC configuration is concerned. Next, we'll configure for our device.

To configure your Edison with openCV, initially execute the following:

```
opkg update
opkg upgrade
```

Finally, after the preceding is successfully executed, run the following:

```
opkg install python-numpy python-opencv
```

This should install all the necessary components. To check whether everything is set up or not, type the following:

```
python
```

And press the *Enter* key. This should enter into the Python shell mode. Next, type the following:

```
importnumpy
import cv2
```

Here is the screenshot of this:

```
COM18 - PuTTY
root@edison:~# python
Python 2.7.3 (default, Jun  6 2016, 13:14:10)
[GCC 4.9.1] on linux2
Type "help", "copyright", "credits" or "license" for more information.
>>> import numpy
>>> import cv2
>>>
```

Python shell

If this doesn't return any error message, then you are all set to go.

 At first we will be covering everything in the PC and after that we'll move on to deploy it to the Intel Edison.

Real-time video display using OpenCV

Before we move on to face detection, let's first see whether we can access our camera or not. To do that, let's write a very simple Python script to display the webcam video feed:

```python
import cv2

cap = cv2.VideoCapture(0)

while(True):
    # Capture frame-by-frame
    ret, frame = cap.read()

    # Our operations on the frame come here
    gray = cv2.cvtColor(frame, cv2.COLOR_BGR2GRAY)

    # Display the resulting frame
    cv2.imshow('frame',gray)
    if cv2.waitKey(1) & 0xFF == ord('q'):
      break
```

```
# When everything done, release the capture
cap.release()
cv2.destroyAllWindows()
```

In the preceding code, we initially import the openCV module as `import cv2`.

Next we initialize the video capture device and set the index to zero as we're using the default webcam that comes with the laptop. For desktop users, you may need to vary the parameter.

After the initialization, in an infinite loop, we read the incoming video frame by frame using `cap.read()`:

```
ret, frame = cap.read()
```

Next we apply some operations on the incoming video feed. Here in the sample, we convert the RGB video frame to a grayscale image:

```
gray = cv2.cvtColor(frame, cv2.COLOR_BGR2GRAY)
```

Finally, the frames are displayed in a separate window:

```
if cv2.waitKey(1) & 0xFF == ord('q'):
break
```

In the preceding two lines, we implement the mechanism of keyboard interrupts. When someone presses *q* or presses the *Esc* key, the display will close.

Once you get the incoming video feed, then we are ready to move to face detection.

Face detection theory

Face detection is a very specific case of object recognition. There are many approaches to face recognition. However, we are going to discuss the two given here:

- Segmentation based on color
- Feature-based recognition

Segmentation based on color

In this technique, the face is segmented out based on skin color. The input of this is typically an RGB image, while in the processing stage we shift it to **Hue saturation value** (**HSV**) or YIQ (Luminance (Y), In-phase Quadrature) color formats. In this process, each pixel is classified as a skin-color pixel or a non-skin-color pixel. The reason behind the use of other color models other than RGB is that sometimes RGB isn't able to distinguish skin colors in different light conditions. This significantly improves while using other color models.

This algorithm won't be used here.

Feature-based recognition

In this technique, we go for certain features and based on that we do the recognition. Use of the haar feature-based cascade for face detection is an effective object detection method proposed by Paul Viola and Michael Jones in their paper "*Rapid Object Detection using a Boosted Cascade of Simple Features*" in 2001. It is a machine learning based approach where a cascade function is trained against a set of positive and negative images. Then it is used to detect objects in other images.

The algorithm initially needs a lot of positive images. In our case, these are images of faces, while negative images which don't contain images of faces. Then we need to extract features from it.

For this purpose, the haar features shown in the following figure are used. Each of the features is a single value obtained by subtracting the sum of pixels under a white rectangle from sum of pixels under a black rectangle:

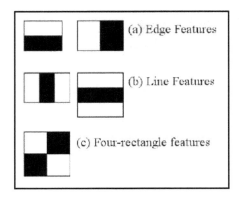

Haar features

The haar classifiers need to be trained for face, eyes, smile, and so on. OpenCV contains a set of predefined classifiers. They are available in the `C:\opencv\build\etc\haarcascades` folder. Now that we know how we can approach face detection, we are going to use the pre-trained haar classifiers for face detection.

Code for face detection

The following is the code for face detection:

```
import cv2
import sys
import os

faceCascade =
cv2.CascadeClassifier('C:/opencv/build/haarcascade_frontalface_default.xml'
)
video_capture = cv2.VideoCapture(0)
while (1):
    # Capture frame-by-frame
    ret, frame = video_capture.read()

    gray = cv2.cvtColor(frame, cv2.COLOR_BGR2GRAY)
    faces = faceCascade.detectMultiScale(gray, 1.3, 5)
    # Draw a rectangle around the faces

    for (x, y, w, h) in faces:
        cv2.rectangle(frame, (x, y), (x+w, y+h), (0, 255, 0), 2)


    # Display the resulting frame
    cv2.imshow('Video', frame)

    if cv2.waitKey(25) == 27:
        video_capture.release()
        break

# When everything is done, release the capture
video_capture.release()
cv2.destroyAllWindows()
```

Let's look at the code line by line:

```
import cv2
import sys
import os
```

Import all the required modules:

```
faceCascade =
cv2.CascadeClassifier('C:/opencv/build/haarcascade_frontalface_default.xml'
)
video_capture = cv2.VideoCapture(0)
```

We select the cascade classifier file. Also we select the video capture device. Make sure you mention the path correctly:

```
ret, frame = video_capture.read()
gray = cv2.cvtColor(frame, cv2.COLOR_BGR2GRAY)
```

In the preceding lines, which are inside the infinite while loop, we read the video frame and convert it from RGB to grayscale:

```
faces = faceCascade.detectMultiScale(gray, 1.3, 5)
```

The preceding line is the most important part of the code. We have actually applied the operation on the incoming feed.

`detectMultiScale` consists of three important parameters. It is a general function for detecting images and since we are applying the face haar cascade, therefore we are detecting faces:

- The first parameter is the input image that needs to be processed. Here we have passed the grayscale version of the original image.
- The second parameter is the scale factor, which provides us with the factor for the creation of a scale pyramid. Typically, around 1.01-1.5 is an appropriate one. The higher the value, the speed increases, but the accuracy decreases.
- The third parameter is `minNeighbours` which affects the quality of the detected regions. A higher value results in less detection. A range of 3-6 is good:

```
# Draw a rectangle around the faces
for (x, y, w, h) in faces:
cv2.rectangle(frame, (x, y), (x+w, y+h), (0, 255, 0), 2)
```

The preceding lines simply draw rectangles around the faces.

Finally, we display the resultant frame and use the keyboard interrupts to release the video capture device and destroy the window.

Now press *F5* to run the code. Initially, it will ask to save the file, and then the execution will begin:

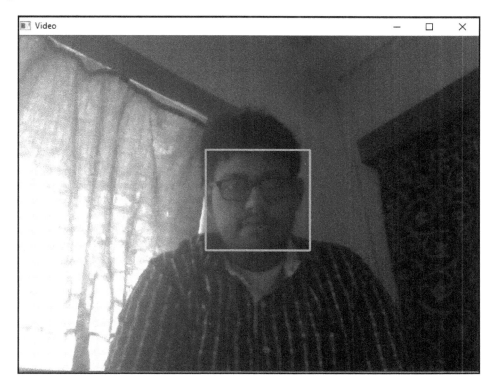

Screenshot of the image window with a face detected

Until now, if everything is carried out in a proper way, you must have a brief idea about face detection and how it can be accomplished using openCV. But now, we need to transfer it to the Intel Edison. Also we need to alter certain parts to meet the capabilities of the device as it doesn't have a display unit and above all it has a RAM of 1 GB.

Intel Edison code

For the Intel Edison, let's find out what is actually possible. We don't have a display, so we can rely only on console messages and LED, perhaps, for visual signals. Next, we may need to optimize the code to run on the Intel Edison. But first let's edit the code discussed previously to include an LED and some kind of messages to the picture:

```
import cv2
import numpy as np
import sys
import os

faceCascade =
cv2.CascadeClassifier('C:/opencv/build/haarcascade_frontalface_default.xml'
)
video_capture = cv2.VideoCapture(0)
led = mraa.Gpio(13)
led.dir(mraa.DIR_OUT)
while (1):
led.write(0)
    # Capture frame-by-frame
    ret, frame = video_capture.read()

    gray = cv2.cvtColor(frame, cv2.COLOR_BGR2GRAY)
    faces = faceCascade.detectMultiScale(gray, 2, 4)
    iflen(faces) > 0:
     print("Detected")
        led.write(1)
    else:
      print("You are clear to proceed")
        led.write(0)
    if cv2.waitKey(25) == 27:
      video_capture.release()
      break

# When everything is done, release the capture
video_capture.release()
cv2.destroyAllWindows()
```

Since the Intel Edison has only one USB port, therefore we have mentioned the parameter of `cv2.VideoCapture` as 0. Also notice the following line:

```
faces = faceCascade.detectMultiScale(gray, 2, 4)
```

You will notice that the parameters have been changed to optimize them for the Intel Edison. You can easily tamper with the parameters to get a good result.

We have included some lines for LED on and off:

```
    ret, frame = video_capture.read()
KeyboardInterrupt
root@edison:~# python faceDetectionEdison.py
Detected
Detected
Detected
Detected
Detected
Detected
Detected
Detected
Detected
Detected
Detected
Detected
Detected
Detected
Detected
Detected
Detected
Detected
Detected
Detected
```

Console output for face detection in images using openCV

This is when you begin to notice that the Intel Edison is simply not meant for image processing because of the RAM.

Now when you are dealing with high processing applications, we cannot rely on the processing power of the Intel Edison alone.

In those cases, we opt for cloud-based solutions. For cloud-based solutions, there are multiple frameworks that exist. One of them is Project Oxford by Microsoft (https://www.microsoft.com/cognitive-services).

Microsoft Cognitive Services provides us with APIs for face detection, recognition, speech recognition, and many more. Use the preceding link to learn more about them.

After all the discussions that we've had in this chapter, we now know that voice recognition performs reasonably well. However, things are not so good with image processing. But why are we focused on using it? The answer lies in that the Intel Edison can definitely be used as an image gathering device while other processing can be carried out on the cloud:

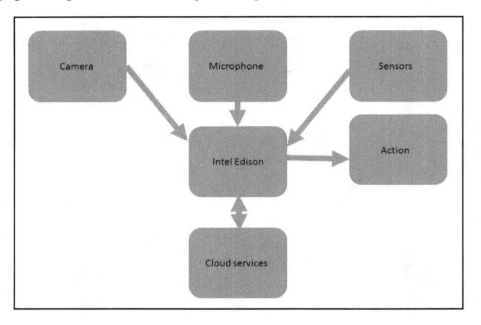

Security-based systems architecture at a glance

Processing can either be performed at the device end or at the cloud end. It all depends on the use case and the availability of resources.

Open-ended task for the reader

The task for this chapter may require a bit of time, but the end result is going to be awesome. Implement Microsoft Cognitive Services to perform facial recognition. Use the Edison to gather data from the user and send it to the service for processing and perform some actions based on the result.

Summary

Throughout this chapter, we have learned some techniques of voice recognition using the Intel Edison. We also learned how image processing can be done in Python and implemented the same on the Intel Edison. Finally, we explored how a real life security-based system would look like and an open-ended question for Microsoft Cognitive Services.

Chapter 5, *Autonomous Robotics with Intel Edison,* will be entirely dedicated to robotics and how the Intel Edison can be used with robotics. We'll be covering both autonomous and manual robotics.

5
Autonomous Robotics with Intel Edison

Robotics is a branch of engineering that deals with the design, development, and application of robots. While robots can take the form of human beings, most robots are designed to perform a specific set of tasks, and it may just look like a machine or anything that is not human. What we are interested in is how the Intel Edison can be used to
develop robots. This chapter will cover the autonomous aspect of robotics and will mainly cover the following topics:

- Architecture of a robotic system
- Intel Edison as a controller
- Connecting sensors with the Intel Edison
- Calibration of sensors with real-time environment
- Actuators: motors, servos, and so on
- Motordrivers: Dual H bridge configuration
- Speed control
- Patching everything together: Line follower robot
- More advanced line follower robots based on the PID control system

This chapter will deal with a line follower robot and will explore all the components of it and discuss some tips and tricks for advanced line following. All the code in this chapter will be written using the Arduino IDE.

Architecture of a typical robotic system

In a typical autonomous robotic system, the **sensor** does the job of gathering data. Based on the data, the controller initially processes it and then performs an action that results in an action by the use of **actuators**:

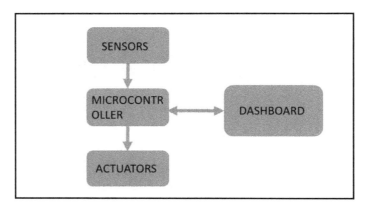

Architecture of a robotic system

We can also have a **dashboard** that may just display what exactly is happening, or sometimes can provide some commands to the robot. Different models exist of the architecture. The preceding architecture is such that the components of a robot are said to be horizontally organized. The information gathered through sensors goes through multiple steps before getting executed as an action. Beside the horizontal flow, there can be vertical flow, where the control may shift to the actuator at any time without completion of the entire process. It's like a queue of cars moving on a highway, and then a single car takes a exit ramp and exits. The line follower robot that we will develop will use the horizontal approach, which is the most primitive.

Intel Edison as a controller

Throughout the book, we have seen the use of the Intel Edison in various applications. In every case, we have stressed on the use of it as a controller because of its features. Now, in the field of robotics, we are again stressing on it as a microcontroller, but the question is, why?

Well, when we are dealing with robotics, in some cases we may need the core capabilities of the Intel Edison. The Intel Edison has a built-in BLE, Wi-Fi, and some features that make it perfect for use in robotics. The Arduino compatible expansion board allows us to use generic Arduino shields for peripherals and motor driving, while the core capabilities provide us some extra functionality for the robot, such as voice-based commands and uploading data to the cloud.

All this comes in a single unit. So, instead of using multiple peripherals, the Intel Edison provides us the flexibility to use everything under a single roof. The ability to attach a camera also adds more spice to it:

Spider bot by Intel. It uses the Intel Edison and a custom expansion board for controlling servos. Picture credit:
`https://i.ytimg.com/vi/3NeJisPvHcU/maxresdefault.jpg`

Connecting sensors to the Intel Edison

In Chapter 2, *Weather Station (IoT)*, we had a brief discussion about sensors. Here we are more focused on sensors that will help the robot find where exactly it is in an environment. We are going to deal with two sensors and how you can calibrate them:

- Ultrasonic sensors (HCSR04)
- Infrared sensors for detection of lines

The reason behind the use of the preceding sensors is that we use these sensors in robotics. However, others are also used, but these are the most commonly used. Now the next question that arises is "how do they operate and how do we hook up with the Intel Edison?"

Let's have a look at each of them.

Ultrasonic sensor (HCSR04)

The main purpose of using this sensor is for the measurement of distances. In our case it may not be needed, but it is very useful in beginner robotics:

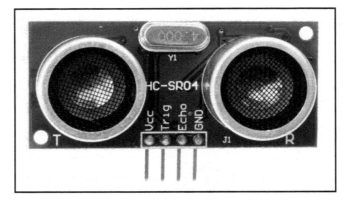

Ultrasonic sensor HCSR04

It emits ultrasonic sound waves, which, if anything is present, then gets back a reflected and picked up by the receiver. Based on the duration of pulse output and the pulse received, we calculate the distance.

The sensor can provide readings ranging from 2 cm to 400 cm. It contains the emitter and the receiver, and thus we can just plug and play with our Intel Edison. The operating voltage is +5V DC. The operation isn't affected by sunlight or dark materials, and is thus efficient for finding distances. Another thing to be noted is that the beam angle is 15 degrees:

Beam angle

Let's have a look at some sample code for using the preceding module.

There is no direct method of calculating the distance directly from the sensors. We need to use two GPIO pins. The first will be the trigger pin, which will be configured as the output, while the second will be the echo pin, which will be in input mode. The moment we send a pulse through the trigger pin, we wait for an incoming pulse in the echo pin. The timing difference and a little bit of mathematics gives us the distance:

```
inttrigPin=2;
intechoPin=3;
void setup()
  {
    pinMode(trigPin,OUTPUT);
    pinMode(echoPin, INPUT);
    Serial.begin(9600);
  }
void loop()
  {
    long duration, distance;
    //Send pulses
    digitalWrite(trigPin, LOW);
    delayMicroseconds(2);
    digitalWrite(trigPin, HIGH);
    delayMicroseconds(10);
    digitalWrite(trigPin, LOW);

    //Receive pulses
    pinMode(echoPin, INPUT);
    duration = pulseIn(echoPin, HIGH);
```

```
    //Calculation
    distance = microsecondsToCentimeters(duration);
    Serial.println(distance);
    delay(200);
}

longmicrosecondsToCentimeters(long microseconds)
    {
    return microseconds / 29 / 2;
    }
```

In the preceding code, concentrate on the loop method. Initially, we send a low pulse to get a clean high pulse. Next, we send a high pulse for 10 seconds. Finally, we get the high pulse in the echo pin and use the `pulseIn()` method to calculate the duration. The duration is sent to another method, where we use the known parameter of the speed of sound to calculate the distance. To use this code on the Edison, connect the HCSR04 sensor to the Edison by following this circuit diagram:

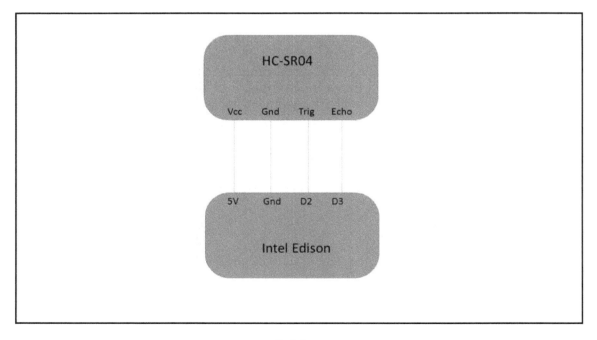

Circuit diagram

Use the preceding circuitry to connect your HCSR04 to your Edison. Next, upload the code and open the serial monitor:

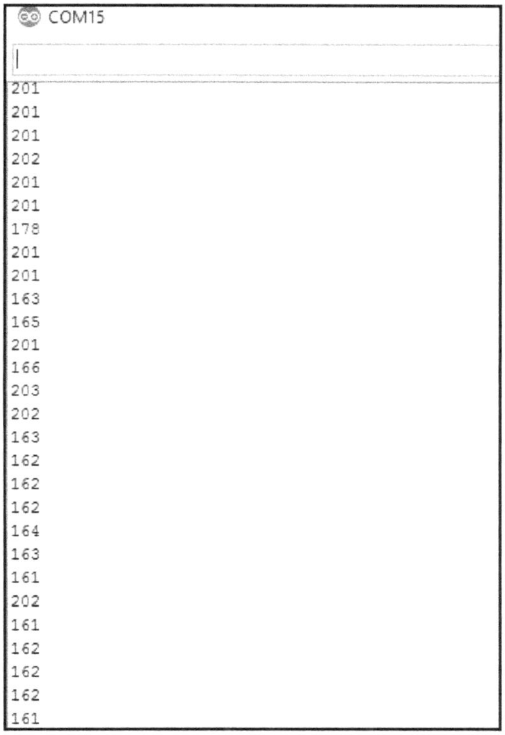

```
COM15

201
201
201
202
201
201
178
201
201
163
165
201
166
203
202
163
162
162
162
164
163
161
202
161
162
162
162
161
```

Serial monitor reading for HCSR04

The preceding screenshot shows us the serial monitor where we get the readings of the distance measured.

Applications of HCSR04

This sensor has lots of applications, especially in autonomous robotics. Even mapping can be accomplished using this sensor. When a HCSR04 is placed on top of a servo motor, the HCSR04 can be used to map the entire 360 degrees. These are extremely useful when we want to perform **simultaneous localization and mapping** (**SLAM**).

Infrared sensors

These sensors have multiple utilities. Starting from line detection to edge detection, these sensors can even be optimized to be used as proximity sensors. Even our smartphones use these sensors. They are usually located near the front speakers. These sensors also work on the principle of sending and receiving signals.

Working methodology

The following image is a commonly used infrared sensor:

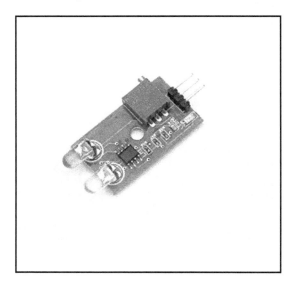

Infrared sensor: Picture source:
http://www.amazon.in/Robosoft-Systems-Single-Sensor-Module/dp/B00U3LKGTG

It has an emitter and a receiver. The output can either be digital or analog. The emitter sends an infrared wave. Based on the object, it is reflected back and the signal is picked up by the receiver. Based on that, we know that we have something in close proximity.

There are multiple applications for infrared sensors:

- Detection of objects in close proximity
- Measuring temperature
- Infrared cameras
- Passive infrared for motion detection

The first was already discussed. The second one is the use of infrared detectors for detection of the infrared spectrum, and based on that, we calculate the temperature. Infrared cameras are widely used in the military, firefighting, and many places where we need to know from a distance that temperatures are high. The third one is the PIR. Passive infrared modules are used for motion detection and are used for automating homes.

Digital and analog outputs for infrared sensors

We have already mentioned that these sensors either give a digital output or an analog output. But why are we so concerned about this? It's mainly because of the reason that in a typical line following we may use digital values, but in high-speed line follower robots, we opt for analog values. The greatest advantage is that there will be a smooth transition from a white surface to a black surface, and in a high-speed line following, we control the speed of the motors based on the sensor analog input, and therefore have more efficient control. For simple purposes, we go for digital output.

Calibration of the infrared sensor module

Infrared sensor modules need to be calibrated for the following reasons:

- The environment may be too bright for the sensor to actually detect anything
- Due to changes in environmental parameters, we may need to calibrate them to suit our needs

Let us consider a line follower robot under two sets of conditions:

- In a normally lit environment (indoor)
- In a sunny environment (outdoor)

When the robot runs efficiently in an indoor environment under ambient light, it is not necessarily true that it will do the same outdoors. The reason is that the sensors are infrared and in a brightly lit environment, especially in sunlight, it's harder to detect. Now, how do we calibrate the sensors? Every sensor module has a potentiometer that controls the sensitivity of the sensor. This potentiometer needs to be adjusted according to the surroundings so that you get appropriate readings.

Now, when you are dealing with robotics, keep some code handy, because it might be required. Similarly for calibration, we follow very simple steps.

Let us consider a simple line follower use case. We need to follow a black line on a white track. The steps are as follows:

1. Place the robot or the sensor in the ideal position (ideal position stands for the position where your robot needs to be, that is, at the center of the line).
2. If the robot is not yet constructed, then cover your sensor with paper and hold it at or about 2 cm above the track.
3. Now check the values of the sensor with the code to be shown in the following steps.
4. There should be a stark difference between the sensor that is above the white line and the sensor that is above the black line. If you have both the sensors above white line then the values should be exactly similar else adjust the potentiometer.
5. Similarly, carry out the process by placing both the sensors over the black line.
6. Carry out the entire process again if the environment or, more precisely, the lighting conditions change.

Hardware setup for calibration and sensor reading

Follow the following circuit diagram for the hardware setup:

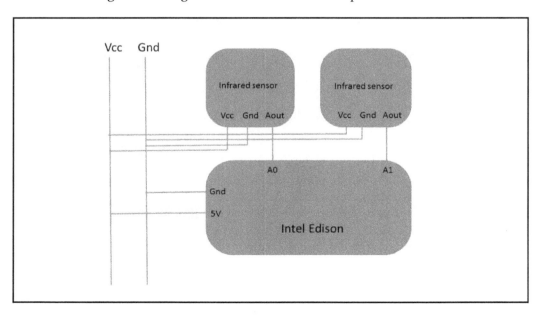

Sensor calibration circuit

The circuit is pretty straightforward and simple to understand. Just attach two sensors to the **A0** and **A1** pins and the common **Vcc** and **Gnd** connections.

Now, let's get on with the code:

```
void setup()
  {
    Serial.begin(9600);
  }
void loop()
  {
    int a= analogRead(A0);
    int b= analogRead(A1);
    Serial.print(a);
    Serial.print("        ");
    Serial.println(b);
    delay(200);
  }
```

The preceding code returns us the values and readings from both the sensors. After you burn the code into the Intel Edison, you will start receiving the values from the sensors. Now, it is required to perform the tests as mentioned in calibration to calibrate the sensors.

Actuators - DC motors and servos

In every robotics project the robot needs to get some mobility. For mobility, we use DC motors or servos. This section will deal with motors and types of motor. The next section will deal with how we can control the motors using the Intel Edison.

Electrical motors are electromechanical devices that convert electrical energy to mechanical energy. The two electrical components in a motor are the field windings and the armature. It's a two pole device. The armature is usually on the rotor, while the field windings are usually on the stator.

Permanent magnet motors use permanent magnets to supply field flux:

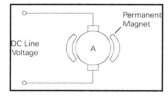

Permanent magnet motor

These motors are restricted by the load they can drive, and that's a disadvantage. However, considering the fact that we are focusing on robotics and on a small-scale basis, we are opting for this kind of motor.

These motors generally operate at 9V-12V, with a maximum full-load current of around 2-3 A. In robotics, we mainly deal with three types of DC motor:

- Continuous DC
- Stepper
- Servo

Continuous DC motors are continuous rotation devices. When the power is switched on the motor rotates, and if polarity is reversed, it rotates in the opposite direction. These motors are widely used for providing mobility to the robot. Normally, DC motors come with gears, which provide some more torque:

A DC motor. Picture source: `http://img.directindustry.com/images_di/photo-g/61070-3667645.jpg`

The next category is a servo motor. The speed of these motors is controlled by varying the width of pulses with a technique known as pulse width modulation. Servo motors are generally an assembly of four things: a DC motor, a gearing set, a control circuit and a position sensor (usually a potentiometer). Generally, a servo motor contains three set of wires: Vcc, Gnd, and signal. Due to thier precise nature, servo motors have a complete different use case, where position is an important parameter. Servo motors do not rotate freely like a standard DC motor. Instead, the angle of rotation is limited to 180 degrees (or so) back and forth. The control signal comes as a **Pulse Width Modulation (PWM)** signal:

A servo motor: Picture

source: https://electrosome.com/wp-content/uploads/2012/06/Servo-Motor.gif

Finally, stepper motors are an advanced form of servo motor. They provide a full 360-degree rotation and it's a continuous motion. The stepper motor utilizes multiple toothed electromagnets arranged together around a central gear to define the position. They require an external controller circuit to individually excite each electromagnet. Stepper motors are available in two varieties; unipolar and bipolar. Bipolar motors are the strongest type of stepper motor and usually have four or eight leads:

Stepper motor

In the project of line following, we'll be dealing with DC motors. But motors cannot be directly hooked up to the Intel Edison, and therefore we need an interfacing circuit, more commonly known as a motor driver.

Motor drivers

Since motors consume voltage and current that cannot be supplied by the GPIO pins alone, we opt for motor drivers. These have an external power supply and use the microcontroller's GPIO pins to receive the control signal. Based on the control signals received, the motor rotates at a particular speed. There are lots of motor drivers available on the market; we will be concentrating on L293D initially, and then a custom and a high-power driver.

The target of any motor driver is to receive control signals from the controller here, the Intel Edison, and send the final output to the motor. Typically, a motor driver can rotate the motor in both directions and also control the speed.

L293D

L293D is a typical motor driver integrated circuit that can drive two motors in both directions. It's like a starter for every robotics project. It's a 16-bit IC:

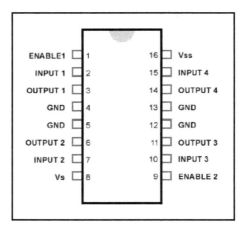

Pinout for L293D. Picture source: http://www.gadgetronicx.com

The maximum voltage for Vs motor supply is 36V. It can supply a max current of 600mA per channel.

It works on the concept of H bridge. The circuit allows the flow of current in either direction. Let's have a look at the H bridge circuit first:

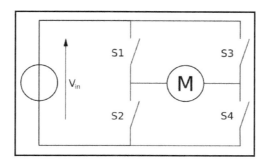

Simple layout of an H bridge circuit. Picture source: https://en.wikipedia.org/

Here, **S1**, **S2**, **S3**, and **S4** are switches that in real life contain transistors. The operation is extremely simple. When **S1** and **S4** are on, the central motor rotates in one direction while the reverse happens when **S2** and **S3** are on. **S1**, **S2**, **S3**, and **S4** receive control signals from the microcontroller and operate the direction of the motor accordingly.

The L293D consists of two such circuits. Thus, we can control up to two motors. One can use the L293D as a module or just as a standalone IC. Normally what we need to worry about are four pins, where we'll send control signals. We have the pin layout for L293D. Let's have a look at what signals will result in what kind of action.

Pins 1-8 are responsible for one motor, while pins 9-16 are responsible for the other.

Enable pin is set to logic high for the operation. The same goes for the other side.

What needs to be tampered with are the input pins 2, 7, 10, and 15. The motors are connected to pins 3, 6, 11, and 14. Vss is for the power supply for the motor, while Vss or Vcc is for the internal power supply. While enable 1 and enable 2, that is pins 1 and 9, are set to high depending on the condition:

Pin 2 or 10	Pin 7 or 15	Motor
High	Low	Clockwise
High	High	Stop
Low	Low	Stop
Low	High	Anti-clockwise

The preceding table summarizes the action triggered based on the input. It is to be noticed that if both the input pins are either on or off then the motor won't rotate. Now in typical L293D modules, only four control pins are exposed and the main voltage supply and Gnd pins are exposed. In the preceding table, it is mentioned as **2 or 10** and **7 or 15**. The pair goes as 2 and 7 or 10 and 15. It means that 2 is on or 10 is off and the same goes for the other as well. The motion of the motor, which is designated as **clockwise** and **anti-clockwise**, depends on the connection of the motor. Assume that the rotation direction reverses when the control signal changes.

Circuit diagram

Construct the following temporary circuit to test out the motors. The enable pins are to be connected to **5V** of the Intel Edison and **Gnd** to **Gnd**.

 It's recommended to use motor driver modules instead of a standalone IC. This allows us to accomplish things in a much simpler way.

CP1 and **CP2** are the control pins for the first motor, while **CP3** and **CP4** are the control pins for the second motor:

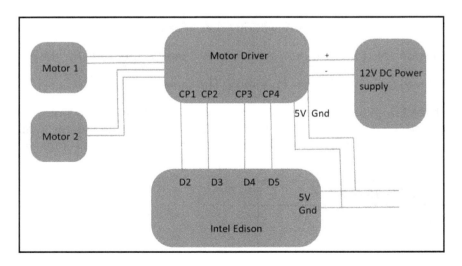

Circuit diagram for motor testing

When dealing with robotics, it is advised to keep sample code with which you can test that your motor driver is working or not. The code discussed here will explain how one can do so.

This part is especially important because it is necessary as the motor testing unit and also for calibration. This code also depends on your motor connection and it may require trial and error:

```
#define M11  2
#define M12  3
#define M21  4
#define M22  5
void setup()
   {
     pinMode(M11,OUTPUT);
     pinMode(M12,OUTPUT);
     pinMode(M21,OUTPUT);
     pinMode(M22,OUTPUT);
   }
void loop()
   {
     forward();
     delay(10000);
     backward();
     delay(10000);
     left();
     delay(10000);
     right();
     delay(10000);
     stop();
     delay(10000);
   }
void forward()
   {
     digitalWrite(M11,HIGH);
     digitalWrite(M12,LOW);
     digitalWrite(M21,HIGH);
     digitalWrite(M22,LOW);
   }
void backward()
   {
     digitalWrite(M11,LOW);
     digitalWrite(M12,HIGH);
     digitalWrite(M21,LOW);
     digitalWrite(M22,HIGH);
   }
```

```
void right()
  {
    digitalWrite(M11,HIGH);
    digitalWrite(M12,LOW);
    digitalWrite(M21,LOW);
    digitalWrite(M22,HIGH);
  }
void left()
  {
    digitalWrite(M11,LOW);
    digitalWrite(M12,HIGH);
    digitalWrite(M21,HIGH);
    digitalWrite(M22,LOW);
  }
void stop()
  {
    digitalWrite(M11,LOW);
    digitalWrite(M12,LOW);
    digitalWrite(M21,LOW);
    digitalWrite(M22,LOW);
  }
```

The preceding code is very simple to understand. Here, we have broken it into several methods for `forward`, `backward`, `left`, `right`, and `stop`. These methods will be responsible for sending control signals to the motor driver using `digitalWrite`. Point to be noted is that this code may not work. It all depends on how your motor is connected. So, when you burn this code to your Intel Edison, make a note of which direction your motor is rotating for the initial 10 seconds. If both motors rotate in the same direction, well then you are lucky enough of not tampering with the connections. If, however, you observe rotation in opposite directions, then reverse the connection of one of the motors from its motor driver. This will allows the motor to rotate in the other direction, so both the motors will rotate in the same direction.

Another important point to be noted is that considering the methods `left` and `right`, we notice that motors rotate in different directions. That's how the steering system of a robot is implemented. It will be discussed in a later section of this chapter.

While dealing with motor drivers, please go through the specifications first. Then go for connections and coding.

Speed control of DC motors

Now that we know how to control a motor and its direction of rotation, let's have a look at controlling the speed of a motor, which is necessary for advanced maneuvering.

The speed control happens through the PWM technique, where we vary the width of pulses to control the speed.

This technique is used to get analog results by digital means. The digital pins on the Intel Edison produce a square wave:

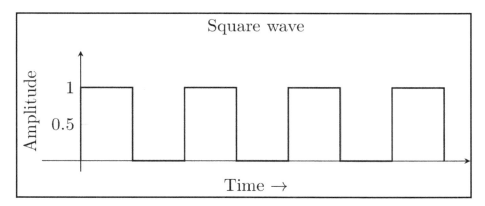

Typical square wave

The on and off pattern can simulate voltages between a full on **5V** and a full off **0V**. This is manipulated by altering the time the signal spends on and the time the signal spends off. The duration of the on time is called **pulse width**. In order for us to get varying pulse values, we change or modulate the pulse width. If this is done fast enough, then the result is a value between 0-5V.

There are PWM pins on the Arduino breakout board for the Intel Edison. These pins are used:

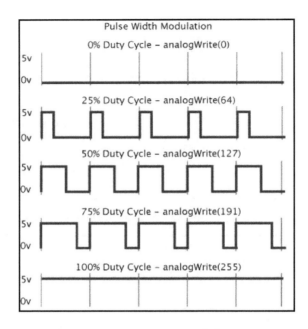

PWM samples. Picture source: www.arduino.cc

Now we can implement this to control the speed of our own motors. In the preceding L293D IC, the enable pins can be used for PWM input.

Modules of L293D mainly expose the following pins:

- Input 1 for motor 1
- Input 2 for motor 1
- Input 1 for motor 2
- Input 2 for motor 2
- Enable for motor 1
- Enable for motor 2
- Vcc
- Gnd

Take a look at the following module:

L293D motor driver module:

A total of eight pins are exposed, as mentioned previously.

Connect the enable pins to any of the PWM pins on the Intel Edison:

- Enable 1 to digital pin 6
- Enable 2 to digital pin 9

Next, to control the speed we need to use the `analogWrite` method in those enabled PWM pins on the Intel Edison.

To set the frequency of the Intel Edison's PWM control, use the example shown in the following link and clone it:

```
https://github.com/MakersTeam/Edison/blob/master/Arduino-Examples/setPWM_Edison
.ino
```

The range of values of the `analogWrite` method is 0-255, where 0 is always off and 255 is always on.

Now using this modify the preceding code to set the enable pin values. An example of using it is shown here. The task of controlling the pins in a fully-fledged motion is left to the reader:

```
#define M11 2
#define M12 3
#define M21 4
#define M22 5
#define PWM 6
void setup()
   {
      pinMode(M11,OUTPUT);
      pinMode(M12,OUTPUT);
      pinMode(M21,OUTPUT);
      pinMode(M22,OUTPUT);
   }
void loop()
   {
      forward();
   }
void forward()
   {
      digitalWrite(M11,HIGH);
      digitalWrite(M12,LOW);
      digitalWrite(M21,HIGH);
      digitalWrite(M22,LOW);
      analogWrite(PWM, 122);
   }
```

In the preceding code, stress on the `forward` method, where we've used `analogWrite (PWM, 122)`. This means that the motor should now rotate half of its original speed. This technique can be used for faster line following robots and speed control.

More advanced motor drivers

While dealing with robotics, there may be some cases where L293D isn't quite a good option due to its current limitation. In those cases, we opt for more powerful drivers. Let's have a look at another product from robokits, which can pretty much drive powerful high torque motors:

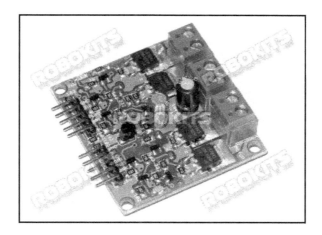

Dual motor driver high power. Picture source:
`http://robokits.co.in/motor-drives/dual-dc-motor-driver-20a`

The preceding motor driver is my personal favorite. It has multiple controls and can drive high torque motors. The driver has the following five control pins:

- **Gnd**: Ground.
- **DIR**: When low, the motor rotates in one direction; when high, it rotates in another direction.
- **PWM**: Pulse width modulation to control the speed of the motor. Recommended frequency range is 20 Hz - 400 Hz.
- **BRK**: When high, it halts the motor in operation.
- **5V**: Regulated 5V output from the motor driver board.

From the description of the pins discussed here, it should be clear as to why this is a better choice.

The voltage and current specifications are as follows:

- Voltage range: 6V to 18V
- Max current: 20 A

The current and the voltage rating help us to drive motors with max load. We have used this motor driver for many of our applications, and it serves without fail. However, there are other motor drivers also on the market that can provide a similar functionality. The choice of which motor driver to use depends on certain factors, as discussed here:

- **Power**: It all depends on how much power the motor needs to run at full capacity. The current drawn at full load and at no load condition. If you are going to use a high torque motor driver with an L293D, you may end up frying your motor driver.
- **Maneuvering**: According to the use case of the problem, the choice of motor is yours and ultimately the choice of the motor driver. In high speed line following, we require PWM capability, thus we need a driver that is capable of handling PWM signals.

Ultimately based on your use case choose your motor driver.

The following is an image of a small yet high performance UGV powered using the previous motor driver that we developed:

Black-e-Track UGV. The UGV's motors are high torque 300 RPM and can climb steep slopes of up to 75 degrees.

Now we have a fairly good idea of how motor drivers work and how we can choose a good motor driver.

Line follower robot (patching everything together)

Based on the previous sections of this chapter, we have got a fairly good idea as to how everything needs to be brought under a single platform. Let's go in to the steps and have a look at how a line follower robot works.

Fundamental concepts of a line follower

The following figure shows the concept of a line follower robot:

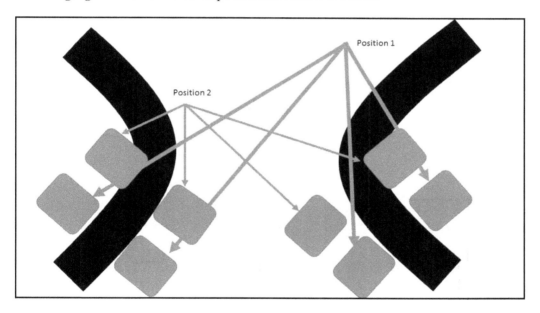

Line follower concept

In the preceding figure, we have two conditions. The first is when the left sensor is over the line and the second is the same for the right sensor. The black line is the track that the robot should follow. There are sensors represented by blocks. Let's consider the left-hand side of the preceding figure.

Initially, both the sensors are over the white surface that is in position 1, as shown in the preceding figure. Next, consider position 2 on the left-hand side of the image. The left sensor comes over the black line and the sensor gets activated. Based on this, we intercept sensor readings and relay control signals to the motor driver. In the preceding case, specific to the left-hand side of the image, we get that the left sensor detected the line and thus the robot should not go forward, instead it should take a slight left turn until and unless both the sensors return the same value.

The same is the case for the right-hand side of the figure where the right sensor detects the black line and the motion execution command is triggered. This is the case of a simple line follower robot where a single colored line needs to be followed. Things get a bit different with multiple colored lines that need to be followed.

Now that we know the exact process of following a line, we can now focus more on how the robot executes turns and the robot structure.

Robot motion execution

To understand how a robot executes its motions, let's consider a four-wheel drive robot:

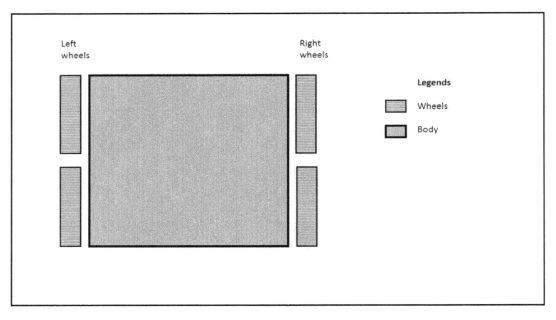

Typical 4WD robot structure

The robot can follow a differential drive-based system where a turn is executed the side where the robot turns, that side's wheel rotates either in the reverse direction, or slows down, or stops. If you are dealing with two wheels, then stopping one side and rotating the other will also do. However, if you are using four wheels, then it's always safe to go for reverse rotation, or it will give rise to wheel slip condition. Again, when using two wheels, we opt for the use of a castor wheel, which keeps the robot balanced. Details of wheel slip condition and other structural elements will be discussed in `Chapter 6`, *Manual Robotics with Intel Edison*.

Hardware requirements for line follower robots

The list of hardware required for a simple two-wheel drive line follower robot is as follows:

- Robot chassis
- 9V DC motors
- Motor driver L293D
- Two infrared sensors
- 9V DC power supply
- Two wheels
- One castor
- Intel Edison, which is used as a controller

The process of attaching the motors to the chassis and the castor won't be shown. A circuit diagram will be shown, and arranging all the components depends on the reader.

Use a two or a four-wheel drive robot chassis, and as we are using two wheels we will fit the castor on the front of the robot. The the sensors should be at the front, on either side of the robot. Let's consider a 2D model of our robot:

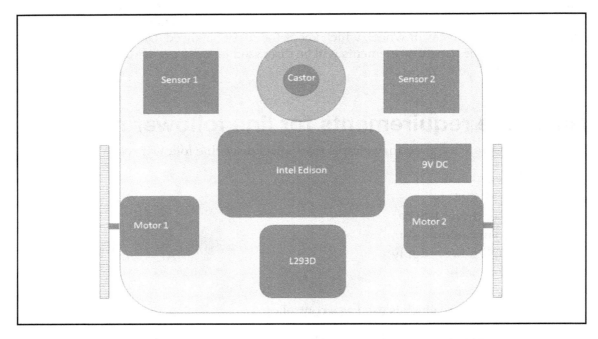

2D robot model of a typical line follower

The preceding figure is a 2D model of the line follower robot with 2WD. The sensors should be on either side and the castor in the middle. While the L293D and the Intel Edison can be located anywhere, the position of the castor, sensors, motors, and obviously the wheels should be the same or similar to the structure shown in the preceding figure.

 Make sure that the distance from the sensor to the ground is optimum for detection of the line. This distance must be calculated during the calibration of the sensors.

The hardware setup usually takes a bit of time as it involves a lot of tickling of wires and loose soldering joints usually add in more problems. Now, before moving forward with the code, let's wire everything up with the following circuit diagram:

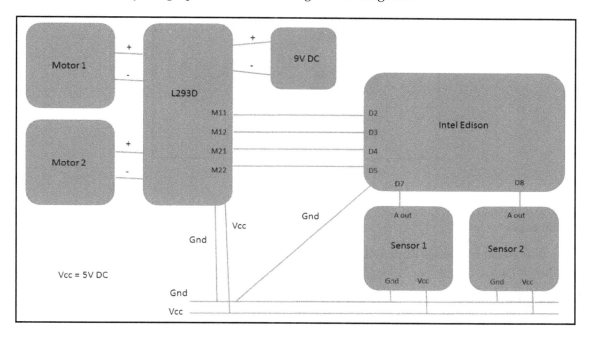

Circuit diagram for line follower robot

The preceding circuit is a combination of what we've done so far in this chapter. A common ground, a **Vcc** line, is created that connects the **L293D**, **Intel Edison**, and the **sensors**. The **9V DC** power supply powers the motors, while the control pins are responsible for sending control signals from the Intel Edison. The output of the sensors is connected to the digital pins of the Intel Edison. Finally, the motor driver controls the motors based on the control signals.

If you have a close look at the preceding circuit diagram, then everything fits into the typical robotics architecture:

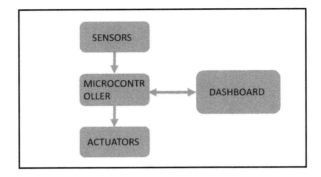

Robotics architecture

But the dashboard is missing. We may add a dashboard, but as of now, we aren't interested in that aspect.

Now that the hardware is done with all the connections and circuitry, let's add a code to it to make it run.

The algorithm is very simple, as follows:

1. Check the left sensor value.
2. If detected, turn left.
3. Or else, check the right sensor value.
4. If detected, turn right.
5. Else if both detected.
6. Stop the motion.
7. Or else, move forward.

The following is the code for this:

```
#define M11 2
#define M12 3
#define M21 4
#define M22 5
#define S1 7
#define S2 8
inta,b;
void setup()
```

```
  {
    pinMode(M11,OUTPUT);
    pinMode(M12,OUTPUT);
    pinMode(M21,OUTPUT);
    pinMode(M22,OUTPUT);
    pinMode(S1,INPUT);
    pinMode(S2,INPUT);
  }
void loop()
  {
    a=digitalRead(S1);
    b=digitalRead(S2);
    //Left turn condition
    if(a==1&&b==0)
      {
        left();
      }
    //Right turn condition
    else if(a==0&&b==1)
      {
        right();
      }
    //Stop condition
    else if(a==1&&b==1)
      {
        stop();
      }
    //By default forward
    else
      {
        forward();
      }
  }
void forward()
  {
    digitalWrite(M11,HIGH);
    digitalWrite(M12,LOW);
    digitalWrite(M21,HIGH);
    digitalWrite(M22,LOW);
  }
void backward()
  {
    digitalWrite(M11,LOW);
    digitalWrite(M12,HIGH);
    digitalWrite(M21,LOW);
    digitalWrite(M22,HIGH);
  }
void right()
```

```
    {
       digitalWrite(M11,HIGH);
       digitalWrite(M12,LOW);
       digitalWrite(M21,LOW);
       digitalWrite(M22,HIGH);
    }
  void left()
    {
       digitalWrite(M11,LOW);
       digitalWrite(M12,HIGH);
       digitalWrite(M21,HIGH);
       digitalWrite(M22,LOW);
    }
  void stop()
    {
       digitalWrite(M11,LOW);
       digitalWrite(M12,LOW);
       digitalWrite(M21,LOW);
       digitalWrite(M22,LOW);
    }
```

In the preceding code, which is very similar to that of the motor testing, only the void loop() is replaced by the main logic, as described in the algorithm. We've used macros for defining sensor and motor pins.

The code initially sets the pins to either input mode or output mode. Next, we store the input values of the sensor. Finally, based on the sensor input, we process the robot motion.

After you burn the code on your Intel Edison, the robot should run. Try a simple track initially, and once your robot runs, then go for a more tight turns. Again, it should be kept in mind that in the preceding code, our right sensor may be your left sensor. In that case you must change the position or just change the condition.

Thus, through a combination of sensors and very simple processing, we can control the motors of a robot and follow a line. Now, if the problem statement asked you to reverse the condition and follow a white line on a black surface, we'd need to tamper with the code a bit, especially in condition checking. The result will be as follows:

```
  void loop()
  {
  int a,b;
    a=digitalRead(S1);
    b=digitalRead(S2);
    //Left turn condition
  if(a==0&&b==1)
    {
```

```
left();
   }
   //Right turn condition
else if(a==1&&b==0)
   {
right();
   }
   //Stop condition
else if(a==0&&b==0)
   {
stop();
   }
   //By default forward
else
   {
forward();
   }
}
```

Just the 1s and 0s need to be interchanged.

Now that we have fairly basic knowledge of developing a basic line follower robot, let's have a brief look at an advanced form of line following and tackle some of the basic concepts.

Advanced line follower robot concepts

So far, we have focused on basic line follower robots of following a single line. Now let's complicate things a bit and try to solve the following section of a track using the previous logic:

Intersection track

In the preceding track, if we use two sensors, then things will get out of hand because the robot should go forward, but according to the algorithm discussed before when both sensors are returning 1, the robot should stop. Then how do we tackle such cases?

The answer lies in the use of more than two sensors. Let's have a look at the following figure:

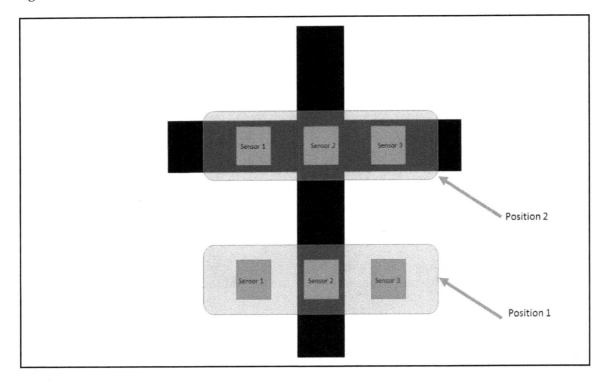

Intersection 3 sensor concept

In the preceding figure, we have shown the use of three sensors. Let's consider the following values:

- If the sensor is on black, it sends 1
- If the sensor is on white, it sends 0

Now, in **position 1**, we get a value of 010 and on **position 1**, the value is 111. This means that 111 represents an intersection. We can have it for left and right junctions too:

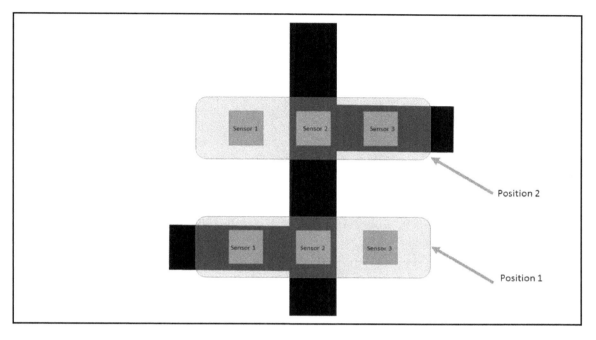

Intersection - 2

Here, the value for left (**position 1**) will be 110 and for right it will be 011. Now it's easier to detect intersecting points and also at the same time prevent the robot from executing false turns. To implement this in code, it's very simple:

```
void loop()
{
int a,b,c;
  a=digitalRead(S1);
  b=digitalRead(S2);
  c=digitalRead(S3);
  //Left turn condition
if(a==1&&b==1&&c==0)
  {
left();
  }
  //Right turn condition
else if(a==0&&b==1&&c==1)
  {
right();
  }
  //Stop condition
else if(a==0&&b==0&&c==0)
```

```
    {
stop();
    }
    //By default forward
else
    {
forward();
    }
}
```

It should be noted that the preceding method needs to be applied based on the scenario of the track. Sometimes it is even required to ignore intersections. It totally depends on the track. The placement of the sensors can also play a very crucial role in line following.

There is a popular line follower robot that has a curved arrangement of sensors. It's the Polulu 3 pi robot:

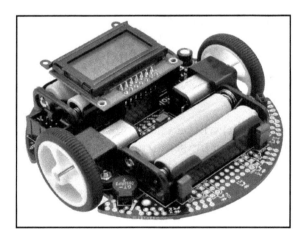

Polulu 3pi robot. Image source: https://www.pololu.com/product/975

Normally we use five or six sensors for a line follower robot. It comes as a module known as a line sensor array:

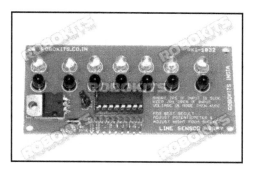

Line sensor array. Image source: `http://robokits.co.in/sensors/line-sensor-array`

Proportional integral derivative - based control

Proportional integral derivative (**PID**) is a control loop feedback mechanism. The main point of using a PID-based control is for efficiently controlling motors or actuators. The main task of the PID is to minimize the error of whatever we are controlling.

It takes an input and then calculates the deviation or error from the intended behavior and ultimately adjusts the output.

Line following may be accurate at lower speeds, but at higher speeds, things may get out of hand or out of control. That's when the PID comes into the picture. Let's have a look at some of the terminology:

- **Error**: The error is something that the device isn't doing the right way. If the RPM of a motor is 380 and the desired RPM is 372 then the error is 380-372=8.
- **Proportional** (**P**): It is directly proportional to the error.
- **Integral** (**I**): It depends on the cumulative error over a period of time.
- **Derivative** (**D**): It depends on the rate of change of error.
- **Constant factor**: When the terms P, I, and D are included in code, it is done by multiplying with some constant factors:
  - P: Factor (Kp)
  - I: Factor (Ki)
  - D: Factor (Kd)

- **Error measurement**: Consider a line follower robot with a five-sensor array that returns digital values; let's have a look at the error measurement. The input obtained from the sensors needs to be weighted depending on the possible combinations of the input. Consider the following table:

Binary value	Weighted value
00001	4
00011	3
00010	2
00110	1
00100	0
01100	-1
01000	-2
11000	-3
10000	-4
00000	-5

The range of values is from -5 to +5. The measurement of the position of the robot is taken several times in a second and then, with the average, we determine the errors.

- **PID formulae**: The error value calculated needs to affect the real motion of the robot. We need to simply add the error value to the output to adjust the robot's motion.

- **Proportional**:

    Difference = Target position - Present position
    *Proportional = Kp * Difference*

    The preceding approach works, but it is found that for a quick response time, if we use a large constant or if the error value is large then the output overshoots the set point. In order to avoid that, the derivative is brought into the picture.

- **Derivative**: Derivative is the rate of change of error.

    Rate of change = (Difference - Previous difference) / Time interval
    *Derivative = Kd * Rate of change*

    The timing interval is obtained by using the timer control of the Intel Edison. This helps us to calculate how quickly the error changes, and based on that, the output is set.

- **Integral:**

    Integral = Integral + Difference
    *Integral = Ki * Integral*

    The integral improves the steady state performance. All the errors are thus added together and the result is applied on the motion of the robot.

    The final control signal is obtained from this:

    Proportional + Integral + Derivative

- **Tuning**: PID implementation can't help you, rather it will degrade the motion of the robot unless and until it is tuned. The tuning parameters are Kp, Ki, and Kd. The tuning value depends on various parameters, such as the friction of the ground, the light conditions, the center of mass, and many more. It varies from one robot to the other.

    Set everything to zero and start with Kp first. Set Kp to 1 and see the condition of the robot. The goal is to make the robot follow the line even if it's wobbly. If the robot overshoots and loses the line, decrease Kp. If it's not able to take turns or seems sluggish, increase Kp.

    Once the robot follows the line more or less correctly, assign 1 to Kd. For now, skip Ki. Increase the value of Ki until you see less wobbling. Once the robot is fairly stable and able to follow the line more or less correctly, assign a value ranging from 0.5-1 in Ki. If the value is too low, not much of a difference would be found. If the value is too high, then the robot may jerk left and right quickly. You may end up incrementing by .01.

PID doesn't implement effective results unless it's properly tuned, so coding only won't yield proper results.

Open-ended question for the reader

The PID use case was explained. Try to implement it in code and write and implement a line follower algorithm with the use of five-six sensors. These practice use cases will explain all the concepts behind line following.

Summary

In this chapter about autonomous robotics, we have covered multiple topics, including dealing from sensors and motor drivers, and how to calibrate sensors and test motor drivers. We also covered line follower robot use cases in detail and also had a chance to look at more advanced controls, and ultimately ended with the PID-based control system.

In Chapter 6, *Manual Robotics with Intel Edison*, we'll cover manual robotics and develop some controller software. We'll also cover more hardware topics pertaining to a robot.

6
Manual Robotics with Intel Edison

In Chapter 5, *Autonomous Robotics with Intel Edison*, we dealt with robotics and the autonomous side of it. Here, we are going to deep dive into the field of manual robotics. A manual robot may not typically be called a robot, so more specifically, we will deal with the manual control of robots that have some autonomous characteristics. We are primarily dealing with the development of UGVs and its control using WPF applications. WPF applications have already been discussed in Chapter 3, *Intel Edison and IoT (Home Automation)*, where we communicated with Edison using the MQTT protocol. Here, we are going to do the same using serial port communication. We will also learn how to make our bot fully wireless. The topics we will be covering are as follows:

- Manual robotic system — architecture and overview
- 2WD and 4WD mechanisms
- Serial port communication with Intel Edison
- Making the bot wireless in robotics
- A simple WPF application to switch an LED on and off using Intel Edison
- High performance motor driver example with code
- Black: e-track platform for UGV
- Universal robot controller for UGV

All the codes for this chapter will be written in Arduino IDE, and for the software side in Visual Studio we are using C# and xaml.

Manual robotic system

We have had a look at the autonomous robotic architecture. Manual robotics also deal with a similar architecture; the only difference being that we have a fully-fledged controller that is responsible for most of the action:

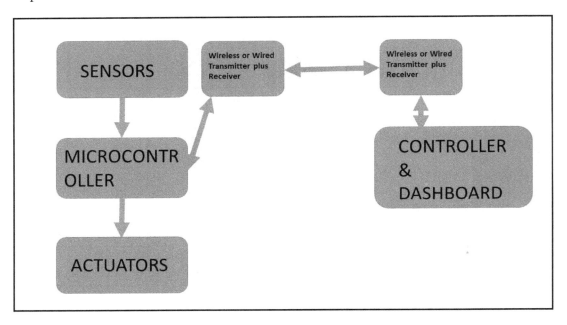

Manual robot architecture

There isn't much of a difference between the architecture discussed here and the one discussed in Chapter 5, *Autonomous Robotics with Intel Edison*. We've added a receiver and a transmitter unit here which would have been present in the earlier use case as well. When dealing with robotics, the entire architecture falls under the same roof.

Manual robotics may not be limited to only manual robots. It may be a blend of manual and autonomous functionality, because a fully manual robot may not typically be called a robot. However, we are aware of **Unmanned Ground Vehicles** (**UGVs**) and **Unmanned Aerial Vehicles** (**UAVs**). Sometimes the terminology may define them as robots, but until, and unless, they don't have at least some manual functionality, they may not be referred to as robots. This chapter mainly deals with UGVs, and like every robot or UGV, we need a sturdy chassis.

Chassis in robotics: 2WD and 4WD

The reader is expected to develop their own robot, and thus you will be required to learn about drive mechanisms and a choice of chassis. Ideally, there are two types of drive mechanisms and the choice for the chassis is done on the basis of the drive mechanism used. Normally we don't want a chassis that over-stresses our motors, nor do we want one that may get stuck while exposed to the outdoor environment. In a typical line follower robot, as discussed in `Chapter 5`, *Autonomous Robotics with Intel Edison*, the most common and the most widely-used drive mechanism is a two-wheel drive, as normally these operate on smooth surfaces and in indoor environments.

Two-wheel drive

Two-wheel drive (**2WD**) refers to the driving mechanism involving two motors and two wheels, and it may typically contain a castor for balancing:

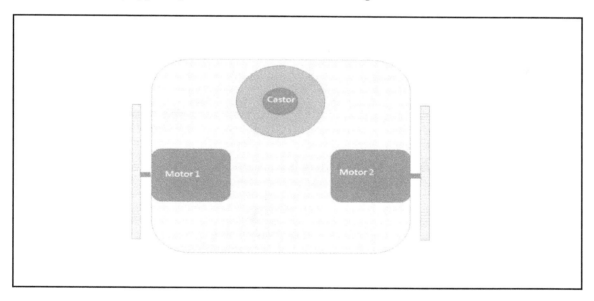

2WD typical layout

The rear motors provide mobility and also acts as a steering mechanism for the robot. For the robot to move right, you may switch off the right motor and let the left motor do the work. However, in that way, the turning radius may be more extreme and the power consumption for the left motor increases as it needs to overcome the force of friction provided by the right wheel. The castor being omni-directional provides less resistance, but this isn't necessarily preferred. The other way is to rotate the right wheel backwards while the left wheel moves forwards; this method allows the robot to turn on its own axis and provide a zero turning radius:

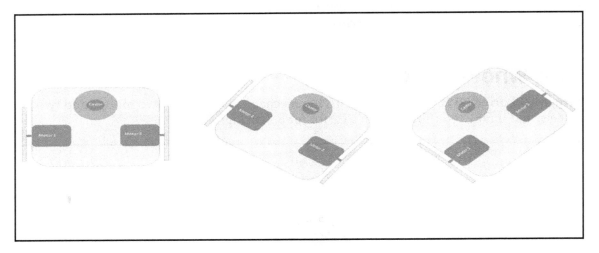

2WD turning of robot on its own axis

When we follow the preceding method, there is a lot less stress on the motors, and with the castor being omni-directional, the robot executes an almost perfect turn.

A chassis can be built with any material and the design should be such that it provides as less stress as possible on the motors.

However, when dealing with four-wheel drives, design plays a factor:

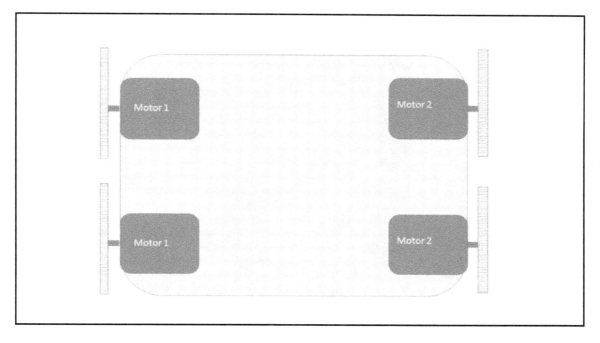

4WD typical drive

Typically, these are powered by four motors (here, motor 1 are the left-side motors, whereas motor 2 are the right-side motors), which can be controlled independently. Usually during rotation, we don't stop the motors on the other side because it creates a lot of pressure on those motors. The other possible option is to rotate on opposite sides—but there is a catch.

Usually, the length of the robot in these cases needs to be either equal to the breadth or even less so. Otherwise, a condition may arise called wheel slip. To prevent such a condition, the design is normally such that the entire model, along with its wheels, fits in a circle, shown as follows:

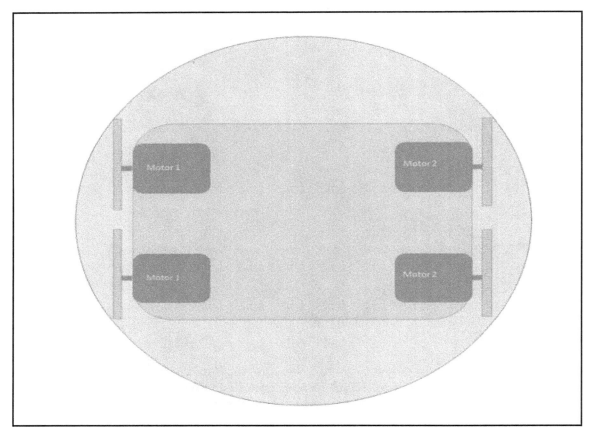

4WD—design

There is another parameter that may be considered and that is the distance between two wheels, as it must be less than the diameter of the wheels. This is because if we are exposed to rough terrain, the bot will be able to come out.

This will happen if the structure of the bot fits in a circle and the length and the distance between the front and rear wheels are less than the distance between the left and right side. Here, while wheel slip happens, it's reduced considerably and is almost negligible. Have a look at the following image for more information:

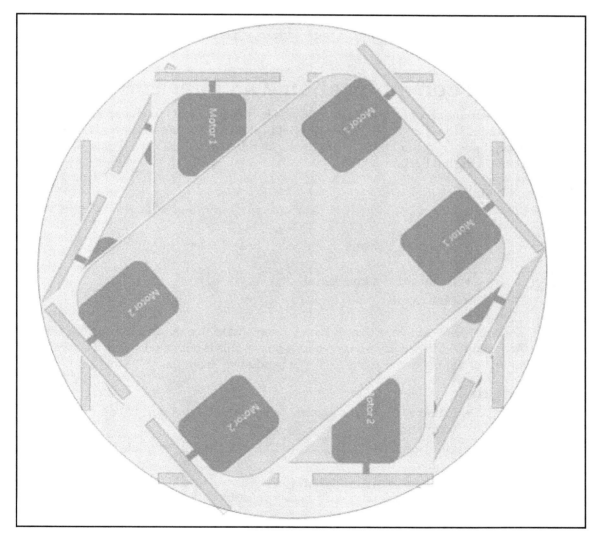

Rotation of a 4WD robot

In the preceding image, the concept should become clearer as the robot tends to stay in the circle it's enclosed in, executing more or less a pivotal turn or a zero radius turn.

Now that we know how the robot chassis can be designed, let's have a look at the ready-made designs available. On sites such as Amazon and eBay, a lot of chassis are available pre-fabricated, following existing design patterns. If you want to fabricate your own chassis, then it's better to follow the preceding design pattern, especially in a 4WD configuration.

Serial port communication with Intel Edison

When we have a manual robot, we need to control it. So, to control it we need some mode of communication. This is attained by using serial port communication. In Intel Edison, we have three serial ports; let's call it Serialx, where x stands for 1 or 2. These serial ports can be accessed by the Arduino IDE:

- **Serial**:
    - **Name**: Multi-gadget, firmware programming, serial console, or OTG port
    - **Location**: USB-micro connector near the center of the Arduino breakout board
    - **ArduinoSWname**: Serial
    - **Linuxname**: `/dev/ttyGS0`

  This port allows us to program Intel Edison and is also the default port for the Arduino IDE. In the Arduino breakout board, this is activated when the toggle switch or SW1 is towards the OTG port and away from the USB slot.

- **Serial1**:
    - **Name**: UART1, the general-purpose TTL-level port (Arduino shield compatibility)
    - **Location**: Pins 0 (RX) and 1 (TX) on the Arduino shield interface headers.
    - **ArduinoSWname**: Serial1
    - **Linuxname**: `/dev/ttyMFD1`

  This port is the pin numbers 0 and 1, which are used as Rx and Tx. This port is used for the remote control of Edison over an RF network or any external Bluetooth device.

- **Serial2**:
    - **Name**: UART2, Linux kernel debug, or debug spew port
    - **Location**: USB-micro connector near the edge of the Arduino board
    - **ArduinoSWname**: Serial2
    - **Linuxname**: /dev/ttyMFD2

This is one of the most useful ports whose communication baud rate is 115200. This is usually the port that is accessed through the PuTTY console and is used to isolate boot problems. When the Serial2 object is created and initialized with Serial2.begin(), the kernel's access to the port is removed and the Arduino sketch is given control until Serial2.end() is invoked.

- **Virtual ports**:
    - **Name**: VCP or virtual communications port (appears only when the Serial-over-USB device is connected)
    - **Location**: Big type A USB port nearest the Arduino power connector
    - **ArduinoSWname**: Not supported by default
    - **Linuxname**: /dev/ttyACMx or /dev/ttyUSBx

This is the USB port of your Intel Edison's Arduino breakout board. The switch must be towards the USB port for enabling the device. Multiple USB devices can be connected using a USB hub.

Consider the following example of code:

```
void setup()
  {
    Serial.begin(9600);
  }

void loop()
  {
    Serial.println("Hi, Reporting in from Intel Edison");
    delay(500);
  }
```

This will just print **Hi, Reporting in from Intel Edison** in the serial monitor. From the code, it's evident that Serial has been used, which is the default one.

Making the system wireless

For making systems wireless in robotics, there are many options available. The choice of hardware and protocol depends on certain factors, which are as follows:

- Availability of mobile network coverage
- Rules and regulations over RF in your operating country
- Maximum distance required
- Availability of Internet connectivity

If we use a GSM module, then mobile network coverage is a must. We may need to get clearance for the RF and ensure that it does not interfere with other signals. The maximum distance is another factor to consider, as distance is limited when using Bluetooth. Bluetooth connectivity can be hampered if the distance exceeds. The same goes for RF, but RF coverage can be increased based on the antenna used. If there is Internet connectivity over an area, then MQTT itself can be used, which was again discussed in `Chapter 3`, *Intel Edison and IoT (Home Automation)*.

RF, or radio frequency, can be used for small applications. Wi-Fi can also be used with Edison, but let's cover a wide spectrum of devices and take a look into how RF can be used.

Normally, RF modules follow a **Universal Asynchronous Receiver Transmitter (UART)** protocol. These generally have a USB link and a serial link. A serial link can be converted with a USB link using a serial to USB converter. There are many options to choose from when buying an RF module set.

 Make a note of what the maximum range and the operating frequency are. All details can be obtained from the place you buy the product.

Normally, the pin out of a RF serial link is shown as follows:

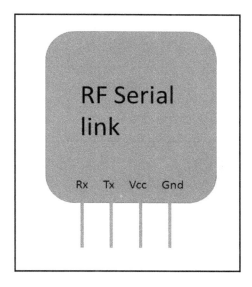

RF serial link pin out

Here is a product of http://robokits.co.in/, which we used in our projects:

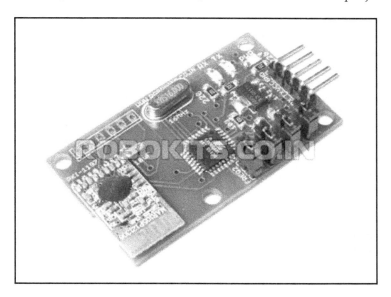

RF USB Serial link. Picture source: http://robokits.co.in/

The module can consist of five pins. We only need to deal with the four pins, as mentioned in the preceding figure.

An RF kit is used to manually control the robot wirelessly by sending commands. These are sent using serial port communication. The controller may use an RF module that has a USB link, or you can use a serial to USB converter to connect it to your PC. The connections of an RF serial link with a serial to USB converter is shown as follows:

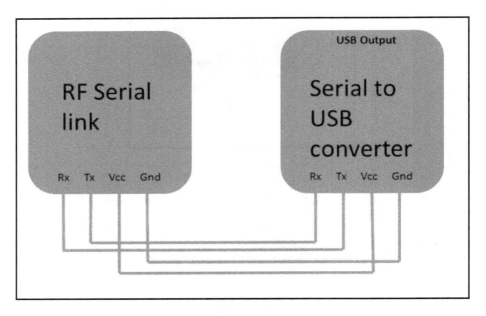

Connections of RF serial link to a serial to USB converter

The connection shown earlier is for connecting an RF serial link to a USB. This applies to the computer side as we want to control it by a PC. We must use two RF modules; one is for Edison and the other is for the controller app or the PC. To connect the RF module to Intel Edison, have a look at the following image:

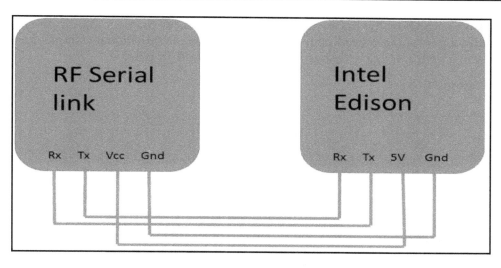

Connections of a RF serial link to Intel Edison

Intel Edison has Rx and Tx pins, which are pins 0 and 1 respectively. The overall architecture is shown as follows:

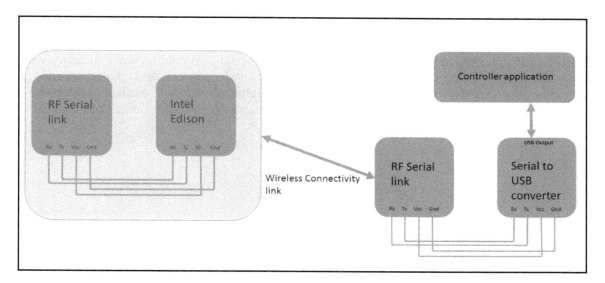

Wireless control of Intel Edison

Now that we know how the hardware pieces are used for wireless communication, the programming part of the preceding model in Intel Edison is ridiculously simple. Just replace Serial with Serial1, as we are using the Rx and Tx pins:

```
void setup()
  {
    Serial1.begin(9600);
  }

void loop()
  {
    Serial1.println("Hi, Reporting in from Intel Edison");
    delay(500);
  }
```

The preceding code sends data to a controller app by using the Rx and Tx pins over an RF network. Now we will have a look on the controller application side, where we will develop a WPF application to control our device.

WPF application for LED on and off

In Chapter 3, *Intel Edison and IoT (Home Automation)*, we looked at using a WPF application and MQTT connection, learning that we could control our Intel Edison using MQTT protocol. However, here, we'll be dealing with serial port communication. Since we have already discussed WPF applications and how to create projects, and created an hello world application, we won't discuss the basics in this chapter, and will instead get into the application directly. Our problem statement in this chapter is to switch an LED on and off using a WPF application via serial port communication.

Start with creating a new WPF project and name it `RobotController`:

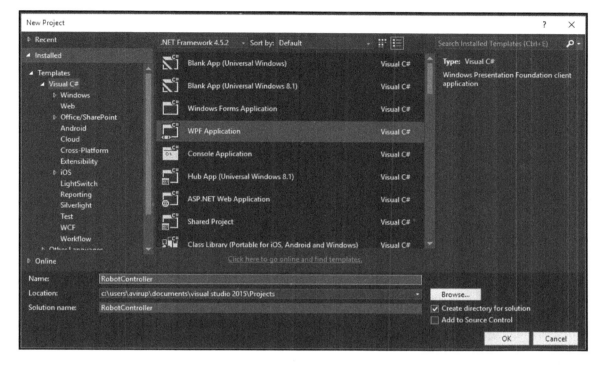

RobotController—1

Next, in **MainWindow.xaml**, we'll design the UI. We'll use the following controls:

- `Buttons`
- `TextBox`
- `TextBlocks`

Design your UI as follows:

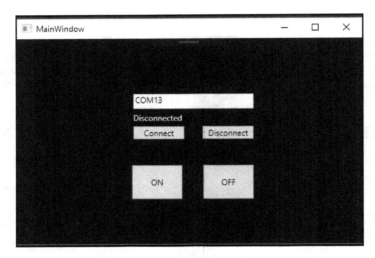

RobotController—2

The xaml code for the preceding UI is as follows:

```xml
<Window x:Class="RobotController.MainWindow"
xmlns="http://schemas.microsoft.com/winfx/2006/xaml/presentation"
xmlns:x="http://schemas.microsoft.com/winfx/2006/xaml"
xmlns:d="http://schemas.microsoft.com/expression/blend/2008"
xmlns:mc="http://schemas.openxmlformats.org/markup-compatibility/2006"
xmlns:local="clr-namespace:RobotController"        mc:Ignorable="d"
Title="MainWindow" Height="350" Width="525" Background="#FF070B64
  <Grid>
    <Grid.ColumnDefinitions>
      <ColumnDefinition Width="335*"/>
      <ColumnDefinition Width="182*"/>
    </Grid.ColumnDefinitions>
    <TextBlock x:Name="status" HorizontalAlignment="Left"
    Margin="172,111,0,0" TextWrapping="Wrap" Text="Disconnected"
    VerticalAlignment="Top" Foreground="White"/>
    <TextBox x:Name="comno" HorizontalAlignment="Left" Height="23"
    Margin="172,83,0,0" TextWrapping="Wrap" Text="COM13"
    VerticalAlignment="Top" Width="177" Grid.ColumnSpan="2"/>
    <Button x:Name="on" Content="ON" HorizontalAlignment="Left"
    Margin="170,191,0,0" VerticalAlignment="Top" Width="74" Height="52"
    Click="on_Click"/>
    <Button x:Name="off" Content="OFF" HorizontalAlignment="Left"
    Margin="275,191,0,0" VerticalAlignment="Top" Width="74" Height="52"
    Grid.ColumnSpan="2" Click="off_Click"/>
```

```
<Button x:Name="connect" Content="Connect"
HorizontalAlignment="Left" Margin="172,132,0,0"
VerticalAlignment="Top" Width="75" Click="connect_Click"/>
<Button x:Name="disconnect" Content="Disconnect"
HorizontalAlignment="Left" Margin="274,132,0,0"
VerticalAlignment="Top" Width="75" Grid.ColumnSpan="2"
Click="disconnect_Click"/>
    </Grid>
</Window>
```

By default, we have written COM13; however, that might change. A total of four buttons are added, which are on, off, connect, and disconnect. We also have a TextBlock to display the status. You can tamper with this code for more customization.

Now our job is to write the backend for this code, which will also include the logic behind it.

Let's first create event handlers. Double click on each of the buttons to create an event. The preceding code contains the event handlers. Once done, include the following namespace for the use of the SerialPort class:

```
using System.IO.Ports;
```

Next, create an object of the SerialPort class:

```
SerialPort sp= new SerialPort();
```

Now navigate to the connect button's event handler method, and here add the code required to connect your app to Intel Edison via a serial port. A try catch block is added to prevent crashes while connecting. The most common reason for a crash is an incorrect port number or the USB is not connected:

```
try
  {
    String portName = comno.Text;
      sp.PortName = portName;
      sp.BaudRate = 9600;
      sp.Open();
      status.Text = "Connected";
  }
    catch (Exception)
  {
     MessageBox.Show("Please give a valid port number or check your
connection");
  }
```

In the preceding code, we stored the `com` port number in a string type variable. Next, we assign the object's `PortName` member with the `portName`. We also set the baud rate to `9600`. Finally, we open the port and write in the status box `connected`.

Next, we write the code for the disconnect event handler:

```
try
  {
    sp.Close();
    status.Text = "Disconnected";
  }
    catch (Exception)
  {
    MessageBox.Show("First Connect and then disconnect");
  }
```

`sp.close()` disconnects the connection. It's safe to write these under a try catch block.

Finally, we write the code for the on and off buttons' event handlers:

```
private void on_Click(object sender, RoutedEventArgs e)
  {
    try
      {
        sp.WriteLine("1");
      }
        catch(Exception)
      {
         MessageBox.Show("Not connected");
      }
  }

private void off_Click(object sender, RoutedEventArgs e)
  {
    try
      {
        sp.WriteLine("2");
      }
        catch (Exception)
      {
        MessageBox.Show("Not connected");
      }
  }
```

In the preceding code, we used the `WriteLine` method and sent a string. The device, which is connected with the application using a serial port, receives the string and an action may be triggered. This sums up the entire process. The entire code for `MainWindow.xaml.cs` is provided as follows:

```
using System;
using System.Collections.Generic;
using System.Linq;
using System.Text;
using System.Threading.Tasks;
using System.Windows;
using System.Windows.Controls;
using System.Windows.Data;
using System.Windows.Documents;
using System.Windows.Input;
using System.Windows.Media;
using System.Windows.Media.Imaging;
using System.Windows.Navigation;
using System.Windows.Shapes;
using System.IO.Ports;

namespace RobotController
{
    /// <summary>
    /// Interaction logic for MainWindow.xaml
    /// </summary>
    public partial class MainWindow : Window
    {
        SerialPort sp = new SerialPort();
        public MainWindow()
        {
            InitializeComponent();
        }

        private void connect_Click(object sender, RoutedEventArgs e)
        {
            try
            {
                String portName = comno.Text;
                sp.PortName = portName;
                sp.BaudRate = 9600;
                sp.Open();
                status.Text = "Connected";
            }
            catch (Exception)
            {
```

```
                    MessageBox.Show("Please give a valid port number or check
your connection");
            }

        }

        private void disconnect_Click(object sender, RoutedEventArgs e)
        {
            try
            {
                sp.Close();
                status.Text = "Disconnected";
            }
            catch (Exception)
            {

                MessageBox.Show("First Connect and then disconnect");
            }
        }

        private void on_Click(object sender, RoutedEventArgs e)
        {
            try
            {
                sp.WriteLine("1");
            }
            catch(Exception)
            {
                MessageBox.Show("Not connected");
            }
        }

        private void off_Click(object sender, RoutedEventArgs e)
        {
            try
            {
                sp.WriteLine("2");
            }
            catch (Exception)
            {
                MessageBox.Show("Not connected");
            }
        }
    }
}
```

Now we have the application ready to control our Intel Edison. Let's test it out. Open up the Arduino IDE. We'll write a small code for Intel Edison that will read serial data from the application so that the on board LED will turn on and off based on the incoming data.

Write the following code to do the same:

```
void setup()
  {
    pinMode(13,OUTPUT);
    Serial.begin(9600);
  }
void loop()
  {
    if(Serial.available()>0)
      {
        char c= Serial.read();
        if(c=='1')
        digitalWrite(13,HIGH);
        else if(c=='2')
        digitalWrite(13,LOW);
      }
  }
```

When you burn this code, go to Visual Studio and run your WPF application. Enter the port number; it must be the same as your Arduino programming port, that is, the serial port. After that, press the on button. The on board LED should glow. It should turn off when you press the off button. Thus, we now have a very basic understanding of how to communicate with Edison using serial port communication via a WPF application. As the chapter progresses, we'll see how to efficiently control a robot with keyboard controls.

High performance motor driver sample with code

In Chapter 5, *Autonomous Robotics with Intel Edison,* we saw an application of L293D and we also wrote some code for it to control motors. However, L293D fails in high performance applications. To tackle this, we had a brief discussion about an alternative high-power driver.

Here, we'll deep dive into the driver, as it has been my personal favorite and is used in virtually all our robots:

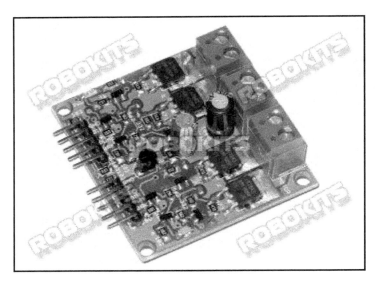

Dual motor driver high power. Picture source:
`http://robokits.co.in/motor-drives/dual-dc-motor-driver-20a`

The driver has the following five control pins:

- **Gnd**: Ground
- **DIR**: When low, the motor rotates in one direction; when high, it rotates in another direction
- **PWM**: Pulse width modulation to control the speed of the motor; the recommended frequency range is 20Hz - 400Hz
- **BRK**: When high, it halts the motor in operation
- **5V**: Regulated 5V output from motor driver board

Now let's write a simple code to operate this driver with all the circuitry:

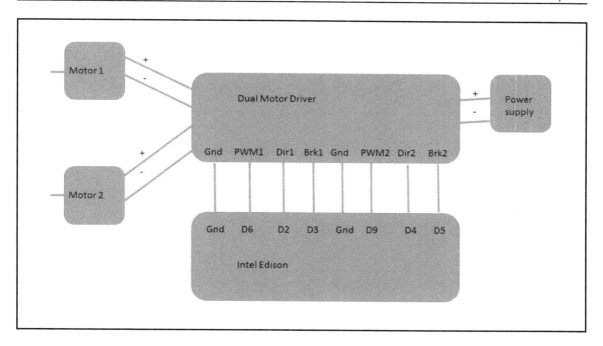

Circuit diagram for motor driver

The preceding circuit is really simple to understand. You don't need to connect the 5V pin. You may use a single ground by shorting two wires of the grounds from the board. Let's now write a code to operate this. This motor driver is very efficient in controlling high torque motors. Since PWM functionality is used, we will therefore use half of the original speed of `122`:

```
int pwm1=6,dir1=2,brk1=3,pwm2=9,dir2=4,brk2=5;
void setup()
{
  pinMode(2,OUTPUT);
  pinMode(6,OUTPUT);
  pinMode(3,OUTPUT);
  pinMode(4,OUTPUT);
  pinMode(5,OUTPUT);
  pinMode(9,OUTPUT);
}
void loop()
{
  forward();
  delay(10000);
  backward();
  delay(10000);
```

```
    left();
    delay(10000);
    right();
    delay(10000);
    stop();
    delay(10000);
}
void forward()
{
  //Left side motors
  digitalWrite(2,LOW); //Direction
  digitalWrite(3,LOW); //Brake
  analogWrite(6,122);
  //Right side motor
  digitalWrite(4,LOW); //Direction
  digitalWrite(5,LOW); //Brake
  analogWrite(9,122);
}
void backward()
{
  //Left side motors
  digitalWrite(2,HIGH);//Direction
  digitalWrite(3,LOW); //Brake
  analogWrite(6,122);
  //Right side motor
  digitalWrite(4,HIGH);//Direction
  digitalWrite(5,LOW); //Brake
  analogWrite(9,122);
}
void left()
{
  //Left side motors
  digitalWrite(2,LOW);//Direction
  digitalWrite(3,LOW); //Brake
  analogWrite(6,122);
  //Right side motor
  digitalWrite(4,HIGH);//Direction
  digitalWrite(5,LOW); //Brake
  analogWrite(9,122);
}
void right()
{
  //Left side motors
  digitalWrite(2,HIGH);//Direction
  digitalWrite(3,LOW); //Brake
  analogWrite(6,122);
  //Right side motor
  digitalWrite(4,LOW);//Direction
```

```
    digitalWrite(5,LOW); //Brake
    analogWrite(9,122);
}
void stop()
{
    //Left side motors
    digitalWrite(2,LOW);//Direction
    digitalWrite(3,HIGH); //Brake
    analogWrite(6,122);
    //Right side motor
    digitalWrite(4,LOW);//Direction
    digitalWrite(5,HIGH); //Brake
    analogWrite(9,122);
}
```

In the preceding code, it is worth noting the functionalities of `Brake` and the `pwm`. Even if you are using a low torque motor, the motor won't rotate if the brake is set to high. Similarly, efficient speed control can be achieved by `pwm` pins. So, by default, we have set everything else on the `pwm` to low. This again depends on the polarity of your motors. Feel free to tamper with the connections so that everything is set with the preceding code. Reverse the connections of your motor if you find an opposite rotation of both sides in the forward condition.

Observe how efficiently motors are controlled by a very simple code.

Now that we know how to control motors more effectively, we'll now move forward with our special black-e-track UGV platform where we developed our own controller for controlling the robot. Almost all the parts were bought from `http://robokits.co.in`.

4WD UGV (black-e-track)

The name might be a bit fancy but this UGV is quite simple with the only difference being that it contained four high torque motors powered by a single dual driver motor driver circuit. Initially, we used two driver circuits but then we shifted to one. It was powered by a Li-Po battery but all tests were conducted using an SMPS. The UGV was controlled by a WPF application with the name of universal remote controller. This UGV was also fitted with a camera with an operating frequency of 5.8 GHz. The UGV was also wireless using a 2.4 GHz RF module. Let's have a look at the hardware required apart from the Intel Edison:

- 30 cm by 30 cm chassis(1)
- 10 cm diameter wheels(4)
- High torque motors 300 RPM 12V(4)

- 20A dual motor driver(1)
- RF 2.4 GHz USB link(1)
- RF 2.4 GHz Serial link(1)
- Li-Po battery (minimum voltage supply: 12V; maximum current drawn: 3-4A)

This section will cover the hardware aspect of it and how to develop the controller application using WPF. The chassis combined with the wheels falls under the deign principle discussed in preceding figure. Let's have a look at the circuit diagram of the UGV. If the robot is made using the earlier mentioned hardware, then the robot will perform well in rough terrain and also be able to climb a steep slope of 60-65 degrees (tested):

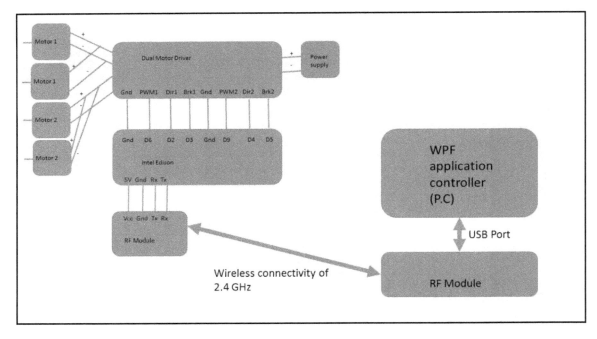

Circuit Diagram for UGV

Motor 1 represents the left hand side motors while motor 2 represents the right hand side motors. Both the left hand side motors are shorted and same goes for the left hand side motors as well. This particular UGV was programmed to receive certain characters through serial port communication and provide some action based on that. Now, let's have a look at the code of the Intel Edison:

```
#define SLOW_SPEED 165
#define MAX_SPEED 255
int pwm1=6,dir1=2,brk1=3,pwm2=9,dir2=4,brk2=5;
```

```
void setup()
{
Serial1.begin(9600);
  pinMode(2,OUTPUT);
  pinMode(6,OUTPUT);
  pinMode(3,OUTPUT);
  pinMode(4,OUTPUT);
  pinMode(5,OUTPUT);
  pinMode(9,OUTPUT);
}
/*
          * 1: Fast front
          * 0: Fast back
          * 3: Fast right
          * 4: Fast left
          * 5: STOP
          * 6: Slow front
          * 7: Slow back
          * 8: Slow right
          * 9: Slow left
          * */
void loop()
{
  if(Serial1.available()>0)
  {
    char c= Serial1.read();
    if(c=='1')
    forward(MAX_SPEED);
    else if(c=='0')
    backward(MAX_SPEED);
    else if(c=='3')
    right(MAX_SPEED);
    else if(c=='4')
    left(MAX_SPEED);
    else if(c=='6')
    forward(SLOW_SPEED);
    else if(c=='7')
    backward(SLOW_SPEED);
    else if(c=='8')
    right(SLOW_SPEED);
    else if(c=='9')
    left(SLOW_SPEED);
    else if(c=='5')
    stop();
  }
}
void forward(int speed)
{
```

```
  //Left side motors
  digitalWrite(2,LOW); //Direction
  digitalWrite(3,LOW); //Brake
  analogWrite(6,speed);
  //Right side motor
  digitalWrite(4,LOW); //Direction
  digitalWrite(5,LOW); //Brake
  analogWrite(9,speed);
}
void backward(int speed)
{
  //Left side motors
  digitalWrite(2,HIGH);//Direction
  digitalWrite(3,LOW); //Brake
  analogWrite(6,speed);
  //Right side motor
  digitalWrite(4,HIGH);//Direction
  digitalWrite(5,LOW); //Brake
  analogWrite(9,speed);
}
void left(int speed)
{
  //Left side motors
  digitalWrite(2,LOW);//Direction
  digitalWrite(3,LOW); //Brake
  analogWrite(6,speed);
  //Right side motor
  digitalWrite(4,HIGH);//Direction
  digitalWrite(5,LOW); //Brake
  analogWrite(9,speed);
}
void right(int speed)
{
  //Left side motors
  digitalWrite(2,HIGH);//Direction
  digitalWrite(3,LOW); //Brake
  analogWrite(6,speed);
  //Right side motor
  digitalWrite(4,LOW);//Direction
  digitalWrite(5,LOW); //Brake
  analogWrite(9,speed);
}
void stop()
{
  //Left side motors
  digitalWrite(2,LOW);//Direction
  digitalWrite(3,HIGH); //Brake
  analogWrite(6,122);
```

```
//Right side motor
digitalWrite(4,LOW);//Direction
digitalWrite(5,HIGH); //Brake
analogWrite(9,122);
}
```

The preceding code executes functions based on the data received. The following table summarises the characters responsible for the data received:

Character Received	Action Undertaken
0	Fast back
1	Fast front
3	Fast right
4	Fast left
5	Stop
6	Slow front
7	Slow back
8	Slow right
9	Slow left

We have created two macros for max and slow speed. The parameters for methods of motion execution is the speed that is passed based on the data received. You can test it using your serial monitor. Now, that we have the hardware lets write a software for it. This software will be able to control the robot using keyboard as well.

Universal robot controller for UGV

Before deep diving into the controller, clone the following GitHub repository to your PC. The code is itself around 350+ lines so some parts are to be discussed:

```
https://github.com/avirup171/bet_controller_urc
```

So initially let's design the UI first:

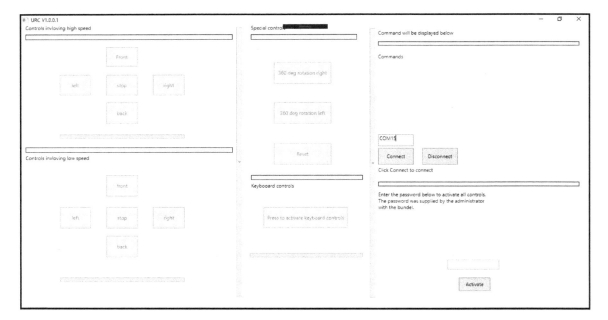

Screenshot of URC

For simplicity two sections of fast and slow controls are included. However it can be merged into one and using a checkbox. We have a connection pane on the right hand top side. The commands are displayed. A default password for 12345 was added which was done to avoid crashes and unauthorized use. However it's a simple controller and can be used with UGVs pretty much efficiently.

If you have a close look over the UI, then you will find a button named **Press to activate keyboard control**. Once you click on the button, the keyboard control gets activated. Now here you need to assign keyboard pressed and keyboard release event. This can be done by selecting the control and in the properties windows, click on the following icon. This manages all the event handlers for the selected control:

Properties window

 When we press the key on a keyboard two events are triggered. The first is when we press the key and the second is when we release the key

Now when you click on it you will get all the possible events associated with it. Scroll down to **KeyDown** and **KeyUp**. Double click on both to create the associated event handlers. The control buttons have a different event associated with them. That is we send data when the button is pressed. When the button is released, **5** which is for stop is sent. You can assign the events by the properties window as shown earlier:

```
<Button x:Name="front_fast" Content="Front" HorizontalAlignment="Left"
VerticalAlignment="Top" Width="75" Margin="204,56,0,0" Height="48"
GotMouseCapture="front_fast_GotMouseCapture"
LostMouseCapture="front_fast_LostMouseCapture" />
```

Assign names to all the buttons and create their respective event handlers. We have also created three progress bars. A xaml code for a progress bar is shown as follows:

```
<ProgressBar x:Name="pb1" HorizontalAlignment="Left" Height="10"
Margin="92,273,0,0" VerticalAlignment="Top" Width="300"/>
```

For the keyboard up and down events, the respective xaml code is:

```
<Button Content="Press to activate keyboard controls" Name="kbp"
HorizontalAlignment="Left" Margin="582,454,0,0" VerticalAlignment="Top"
Width="202" Height="52" KeyDown="kbp_KeyDown" KeyUp="Kbp_KeyUp"/>
```

Two events are created. One for the key pressed and another for key released.

The xaml code for the preceding UI is given as follows:

```
<Window x:Class="BET_Controller_v2.MainWindow"
xmlns="http://schemas.microsoft.com/winfx/2006/xaml/presentation"
xmlns:x="http://schemas.microsoft.com/winfx/2006/xaml"        Title="URC
V1.0.0.1" Height="720" Width="1360" Loaded="Window_Loaded">
  <Grid>
    <TextBlock HorizontalAlignment="Left" TextWrapping="Wrap"
    Text="Controls invloving high speed" VerticalAlignment="Top"
    Height="25" Width="269" Margin="10,0,0,0"/>
    <Rectangle Fill="#FFF4F4F5" HorizontalAlignment="Left" Height="13"
    Stroke="Black" VerticalAlignment="Top" Width="498"
    Margin="10,25,0,0"/>
      <Button x:Name="front_fast" Content="Front"
      HorizontalAlignment="Left" VerticalAlignment="Top" Width="75"
      Margin="204,56,0,0" Height="48"
      GotMouseCapture="front_fast_GotMouseCapture"
      LostMouseCapture="front_fast_LostMouseCapture" />
      <Button x:Name="back_fast" Content="back"
      HorizontalAlignment="Left" VerticalAlignment="Top" Width="75"
      Margin="204,193,0,0" Height="53"
      GotMouseCapture="back_fast_GotMouseCapture"
      LostMouseCapture="back_fast_LostMouseCapture"/>
      <Button x:Name="left_fast" Content="left"
      HorizontalAlignment="Left" VerticalAlignment="Top" Width="74"
      Margin="92,125,0,0" Height="50"
      GotMouseCapture="left_fast_GotMouseCapture"
      LostMouseCapture="left_fast_LostMouseCapture" />
      <Button x:Name="right_fast" Content="right"
      HorizontalAlignment="Left" VerticalAlignment="Top" Width="76"
      Margin="316,125,0,0" Height="50"
      GotMouseCapture="right_fast_GotMouseCapture"
      LostMouseCapture="right_fast_LostMouseCapture" />
      <Button x:Name="stop_fast" Content="stop"
      HorizontalAlignment="Left" VerticalAlignment="Top" Width="75"
      Margin="204,125,0,0" Height="50"
      RenderTransformOrigin="0.362,0.5" Click="stop_fast_Click" />
```

```
   <Rectangle Fill="#FFF4F4F5" HorizontalAlignment="Left" Height="13"
   Stroke="Black" VerticalAlignment="Top" Width="498"
   Margin="10,305,0,0"/>
<TextBlock HorizontalAlignment="Left" TextWrapping="Wrap"
Text="Controls involving low speed" VerticalAlignment="Top"
Height="25" Width="269" Margin="10,323,0,0"/>
   <Button x:Name="front_slow" Content="front"
   HorizontalAlignment="Left" Margin="204,374,0,0"
   VerticalAlignment="Top" Width="75" Height="53"
   GotMouseCapture="front_slow_GotMouseCapture"
   LostMouseCapture="front_slow_LostMouseCapture"/>
   <Button x:Name="back_slow" Content="back"
   HorizontalAlignment="Left" Margin="204,526,0,0"
   VerticalAlignment="Top" Width="75" Height="53"
   LostMouseCapture="back_slow_LostMouseCapture"
   GotMouseCapture="back_slow_GotMouseCapture" />
   <Button x:Name="right_slow" Content="right"
   HorizontalAlignment="Left" Margin="316,454,0,0"
   VerticalAlignment="Top" Width="76" Height="53"
   LostMouseCapture="right_slow_LostMouseCapture"
   GotMouseCapture="right_slow_GotMouseCapture" />
   <Button x:Name="left_slow" Content="left"
   HorizontalAlignment="Left" Margin="92,454,0,0"
   VerticalAlignment="Top" Width="74" Height="53"
   GotMouseCapture="left_slow_GotMouseCapture"
   LostMouseCapture="left_slow_LostMouseCapture" />
   <Button x:Name="stop_slow" Content="stop"
   HorizontalAlignment="Left" Margin="204,454,0,0"
   VerticalAlignment="Top" Width="75" Height="53"
   Click="stop_slow_Click"/>
 <Rectangle Fill="#FFF4F4F5" HorizontalAlignment="Left" Height="13"
 Stroke="Black" VerticalAlignment="Top" Width="255"
 Margin="552,25,0,0"/>
<TextBlock HorizontalAlignment="Left" TextWrapping="Wrap"
Text="Special controls" VerticalAlignment="Top" Height="25"
Width="269" Margin="552,0,0,0"/>
   <Button x:Name="rr360" Content="360 deg rotation right"
   HorizontalAlignment="Left" Margin="607,94,0,0"
   VerticalAlignment="Top" Width="138" Height="53"
   RenderTransformOrigin="0.498,0.524" Click="rr360_Click"/>
   <Button x:Name="lr360" Content="360 deg rotation left"
   HorizontalAlignment="Left" Margin="607,193,0,0"
   VerticalAlignment="Top" Width="138" Height="53"
   RenderTransformOrigin="0.498,0.524" Click="lr360_Click" />
<TextBlock HorizontalAlignment="Left" TextWrapping="Wrap"
Text="Command will be displayed below" VerticalAlignment="Top"
Height="25" Width="484" Margin="858,13,0,0"/>
 <Rectangle Fill="#FFF4F4F5" HorizontalAlignment="Left" Height="10"
```

```
    Stroke="Black" VerticalAlignment="Top" Width="484"
    Margin="858,43,0,0"/>
<TextBlock x:Name="cmd" HorizontalAlignment="Left"
Margin="858,72,0,0" TextWrapping="Wrap" Text="Commands"
VerticalAlignment="Top" Height="174" Width="484"/>
<TextBox x:Name="cno" HorizontalAlignment="Left" Height="29"
Margin="858,273,0,0" TextWrapping="Wrap" Text="COM13"
VerticalAlignment="Top" Width="85"/>
    <Button x:Name="connect" Content="Connect"
    HorizontalAlignment="Left" Margin="858,307,0,0"
    VerticalAlignment="Top" Width="85" Height="41"
    RenderTransformOrigin="0.498,0.524" Click="connect_Click" />
    <Button x:Name="disconnect" Content="Disconnect"
    HorizontalAlignment="Left" Margin="964,307,0,0"
    VerticalAlignment="Top" Width="85" Height="41"
    RenderTransformOrigin="0.498,0.524" Click="disconnect_Click" />
    <Button x:Name="rst" Content="Reset" HorizontalAlignment="Left"
    Margin="607,295,0,0" VerticalAlignment="Top" Width="138"
    Height="53" RenderTransformOrigin="0.498,0.524" Click="rst_Click"
    />
<TextBlock HorizontalAlignment="Left" Margin="858,353,0,0" x:Name="s"
TextWrapping="Wrap" Text="Click Connect to connect"
VerticalAlignment="Top" Height="33" Width="191"/>
  <ProgressBar x:Name="pbkc" HorizontalAlignment="Left" Height="10"
  Margin="549,569,0,0" VerticalAlignment="Top" Width="272"/>
  <ProgressBar x:Name="pb2" HorizontalAlignment="Left" Height="10"
  Margin="92,631,0,0" VerticalAlignment="Top" Width="300"/>
  <ProgressBar x:Name="pb1" HorizontalAlignment="Left" Height="10"
  Margin="92,273,0,0" VerticalAlignment="Top" Width="300"/>
  <Rectangle Fill="#FFF4F4F5" HorizontalAlignment="Left" Height="12"
  Stroke="Black" VerticalAlignment="Top" Width="269"
  Margin="552,374,0,0"/>
<TextBlock HorizontalAlignment="Left" TextWrapping="Wrap"
Text="Keybooard controls" VerticalAlignment="Top" Height="25"
Width="269" Margin="552,392,0,0"/>
    <Button Content="Press to activate keyboard controls" Name="kbp"
    HorizontalAlignment="Left" Margin="582,454,0,0"
    VerticalAlignment="Top" Width="202" Height="52"
    KeyDown="kbp_KeyDown" KeyUp="Kbp_KeyUp"/>
  <Rectangle Fill="#FFF4F4F5" HorizontalAlignment="Left" Height="11"
  Stroke="Black" VerticalAlignment="Top" Width="484"
  Margin="858,391,0,0"/>
<TextBlock HorizontalAlignment="Left" TextWrapping="Wrap" Text="Enter
the password below to activate all controls. The password was
supplied by the administrator with the bundel."
VerticalAlignment="Top" Height="51" Width="269"
Margin="858,414,0,0"/>
    <Button x:Name="activate" Content="Activate"
```

```
        HorizontalAlignment="Left" Margin="1052,631,0,0"
        VerticalAlignment="Top" Width="75" Height="33"
        Click="activate_Click" KeyDown="activate_KeyDown"/>
    <PasswordBox x:Name="pswrdbox" HorizontalAlignment="Left"
  Margin="1025,585,0,0" VerticalAlignment="Top" Height="25" Width="124"
  PasswordChar="*" FontSize="24"/>
        <Rectangle Fill="#FFF4F4F5" HorizontalAlignment="Left" Height="645"
        Stroke="Black" VerticalAlignment="Top" Width="5" Margin="834,8,0,0"
        RenderTransformOrigin="0.5,0.5">
        </Rectangle><Rectangle Fill="#FFF4F4F5" HorizontalAlignment="Left"
        Height="645" Stroke="Black" VerticalAlignment="Top" Width="5"
        Margin="539,19,0,0" RenderTransformOrigin="0.5,0.5"/>
    </Grid>
</Window>
```

Now that the UI is ready, let's go to the main C# code. The event handlers are also in place. Initially include the `System.IO.Ports` namespace and create an object of that class. After that the keyboard pressed event will be handled with our code:

```
private void kbp_KeyDown(object sender, KeyEventArgs e)
        {
            Keyboard.Focus(kbp);
            if (e.Key == Key.W)
            {
                sp.WriteLine("1");
                pbkc.IsIndeterminate = true;
                cmd.Text = "W: fast forward";
            }
            else if (e.Key == Key.S)
            {
                sp.WriteLine("0");
                pbkc.IsIndeterminate = true;
                cmd.Text = "S: fast back";
            }
            else if (e.Key == Key.A)
            {
                sp.WriteLine("4");
                pbkc.IsIndeterminate = true;
                cmd.Text = "A: fast left";
            }
            else if (e.Key == Key.D)
            {
                sp.WriteLine("3");
                pbkc.IsIndeterminate = true;
                cmd.Text = "D: fast right";
            }
            else if (e.Key == Key.NumPad8)
            {
```

```
            sp.WriteLine("6");
            pbkc.IsIndeterminate = true;
            cmd.Text = "9: slow front";
        }
        else if (e.Key == Key.NumPad2)
        {
            sp.WriteLine("7");
            pbkc.IsIndeterminate = true;
            cmd.Text = "2: slow back";
        }
        else if (e.Key == Key.NumPad6)
        {
            sp.WriteLine("8");
            pbkc.IsIndeterminate = true;
            cmd.Text = "6: slow right";
        }
        else if (e.Key == Key.NumPad4)
        {
            sp.WriteLine("9");
            pbkc.IsIndeterminate = true;
            cmd.Text = "D: slow left";
        }

    }
```

In the preceding code, we used the following keys:

Serial No	Keyboard keys	Commands executed
1	W	Fast forward
2	A	Fast left turn
3	S	Fast backward
4	D	Fast right turn
5	Numpad 8	Slow forward
6	Numpad 2	Slow Backward
7	Numpad 4	Slow left turn
8	Numpad 6	Slow right turn

Based on the input, we sent that particular character. While for the key up or key released event, we simply send 5 which means stop:

```
private void Kbp_KeyUp(object sender, KeyEventArgs e)
  {
    sp.WriteLine("5");
    pbkc.IsIndeterminate = false;
    cmd.Text = "STOP";
  }
```

The connect and disconnect events are same as before. Now each button will have two methods. The first one is of GotMouseCapture and the second one is of LostMouseCapture.

Take the example of the front button under fast control:

```
private void front_fast_GotMouseCapture(object sender, MouseEventArgs e)
  {
    sp.WriteLine("1");
    cmd.Text = "Fast Forward";
    pb1.IsIndeterminate = true;
  }

private void front_fast_LostMouseCapture(object sender, MouseEventArgs e)
  {
    sp.WriteLine("5");
    cmd.Text = "STOP";
    pb1.IsIndeterminate = false;
  }
```

Similarly apply it for the other controls. Only 360 degree left and right is associated with a button click event. The entire code is pasted as shown below of MainWindow.xaml.cs:

```
using System;
using System.Collections.Generic;
using System.Linq;
using System.Text;
using System.Threading.Tasks;
using System.Windows;
using System.Windows.Controls;
using System.Windows.Data;
using System.Windows.Documents;
using System.Windows.Input;
using System.Windows.Media;
using System.Windows.Media.Imaging;
using System.Windows.Navigation;
using System.Windows.Shapes;
```

```
using System.IO.Ports;

namespace BET_Controller_v2
{
    /// <summary>
    /// Interaction logic for MainWindow.xaml
    /// </summary>
    public partial class MainWindow : Window
    {
        SerialPort sp = new SerialPort();
        public MainWindow()
        {
            InitializeComponent();
            Closing += new
System.ComponentModel.CancelEventHandler(MainWindow_Closing);
        }

        private void MainWindow_Closing(object sender,
System.ComponentModel.CancelEventArgs e)
        {
            if (MessageBox.Show("Do you really want to exit?", "Exit",
MessageBoxButton.YesNo) == MessageBoxResult.No)
            {
                e.Cancel = true;
            }
        }
/*
        * 1: Fast front
        * 0: Fast back
        * 3: Fast right
        * 4: Fast left
        * 5: STOP
        * 6: Slow front
        * 7: Slow back
        * 8: Slow right
        * 9: Slow left
        * */
        //Keyboard Controls
        private void kbp_KeyDown(object sender, KeyEventArgs e)
        {
            Keyboard.Focus(kbp);
            if (e.Key == Key.W)
            {
                sp.WriteLine("1");
                pbkc.IsIndeterminate = true;
                cmd.Text = "W: fast forward";
            }
            else if (e.Key == Key.S)
```

```
                {
                    sp.WriteLine("0");
                    pbkc.IsIndeterminate = true;
                    cmd.Text = "S: fast back";
                }
                else if (e.Key == Key.A)
                {
                    sp.WriteLine("4");
                    pbkc.IsIndeterminate = true;
                    cmd.Text = "A: fast left";
                }
                else if (e.Key == Key.D)
                {
                    sp.WriteLine("3");
                    pbkc.IsIndeterminate = true;
                    cmd.Text = "D: fast right";
                }
                else if (e.Key == Key.NumPad8)
                {
                    sp.WriteLine("6");
                    pbkc.IsIndeterminate = true;
                    cmd.Text = "9: slow front";
                }
                else if (e.Key == Key.NumPad2)
                {
                    sp.WriteLine("7");
                    pbkc.IsIndeterminate = true;
                    cmd.Text = "2: slow back";
                }
                else if (e.Key == Key.NumPad6)
                {
                    sp.WriteLine("8");
                    pbkc.IsIndeterminate = true;
                    cmd.Text = "6: slow right";
                }
                else if (e.Key == Key.NumPad4)
                {
                    sp.WriteLine("9");
                    pbkc.IsIndeterminate = true;
                    cmd.Text = "D: slow left";
                }

            }
    //Key release event handlers
            private void Kbp_KeyUp(object sender, KeyEventArgs e)
            {
                sp.WriteLine("5");
                pbkc.IsIndeterminate = false;
```

```
                        cmd.Text = "STOP";
            }
//Connect to the hardware
        private void connect_Click(object sender, RoutedEventArgs e)
        {
            try
            {
                String pno = cno.Text;
                sp.PortName = pno;
                sp.BaudRate = 9600;
                sp.Open();
                s.Text = "Connected";
            }
            catch (Exception)
            {

                MessageBox.Show("Please check the com port number or the
hardware attached to it");
            }
        }
        //Disconnect from the hardware
        private void disconnect_Click(object sender, RoutedEventArgs e)
        {
            try
            {
                sp.Close();
                s.Text = "Disconnected";
            }
            catch (Exception)
            {

                MessageBox.Show("Some error occured with the connection");
            }
        }

        private void front_fast_GotMouseCapture(object sender,
MouseEventArgs e)
        {
            sp.WriteLine("1");
            cmd.Text = "Fast Forward";
            pb1.IsIndeterminate = true;
        }

        private void front_fast_LostMouseCapture(object sender,
MouseEventArgs e)
        {
            sp.WriteLine("5");
            cmd.Text = "STOP";
```

```
            pb1.IsIndeterminate = false;
        }

        private void back_fast_GotMouseCapture(object sender,
MouseEventArgs e)
        {
            sp.WriteLine("0");
            cmd.Text = "Fast Backward";
            pb1.IsIndeterminate = true;
        }

        private void back_fast_LostMouseCapture(object sender,
MouseEventArgs e)
        {
            sp.WriteLine("5");
            cmd.Text = "STOP";
            pb1.IsIndeterminate = false;
        }

        private void left_fast_GotMouseCapture(object sender,
MouseEventArgs e)
        {
            sp.WriteLine("4");
            cmd.Text = "Fast left";
            pb1.IsIndeterminate = true;
        }

        private void left_fast_LostMouseCapture(object sender,
MouseEventArgs e)
        {
            sp.WriteLine("5");
            cmd.Text = "STOP";
            pb1.IsIndeterminate = false;
        }

        private void activate_Click(object sender, RoutedEventArgs e)
        {
         // Password and activation section
            string s = pswrdbox.Password;
            if(s=="12345")
            {
                MessageBox.Show("Congrats Black e Track Controller V2 is
activated");
                front_fast.IsEnabled = true;
                back_fast.IsEnabled = true;
                stop_fast.IsEnabled = true;
                left_fast.IsEnabled = true;
                right_fast.IsEnabled = true;
```

```
            front_slow.IsEnabled = true;
            back_slow.IsEnabled = true;
            right_slow.IsEnabled = true;
            left_slow.IsEnabled = true;
            stop_slow.IsEnabled = true;
            rr360.IsEnabled = true;
            lr360.IsEnabled = true;
            rst.IsEnabled = true;
            kbp.IsEnabled = true;
        }
        else
        {
            MessageBox.Show("Sorry you have entered wrong password.
Please enter the correct credentials or contact your system
administrator.");
        }
    }

    private void right_fast_GotMouseCapture(object sender,
MouseEventArgs e)
    {
        sp.WriteLine("3");
        cmd.Text = "Fast Right";
        pb1.IsIndeterminate = true;
    }

    private void right_fast_LostMouseCapture(object sender,
MouseEventArgs e)
    {
        sp.WriteLine("5");
        cmd.Text = "STOP";
        pb1.IsIndeterminate = false;
    }

    private void front_slow_GotMouseCapture(object sender,
MouseEventArgs e)
    {
        sp.WriteLine("6");
        cmd.Text = "Slow Front";
        pb2.IsIndeterminate = true;
    }


    private void front_slow_LostMouseCapture(object sender,
MouseEventArgs e)
    {
        sp.WriteLine("5");
        cmd.Text = "STOP";
```

```
            pb2.IsIndeterminate = false;
        }
        private void back_slow_LostMouseCapture(object sender,
MouseEventArgs e)
        {
            sp.WriteLine("5");
            cmd.Text = "STOP";
            pb2.IsIndeterminate = false;
        }


        private void back_slow_GotMouseCapture(object sender,
MouseEventArgs e)
        {
            sp.WriteLine("7");
            cmd.Text = "Slow Back";
            pb2.IsIndeterminate = true;
        }
        private void left_slow_GotMouseCapture(object sender,
MouseEventArgs e)
        {
            sp.WriteLine("9");
            cmd.Text = "Slow Left";
            pb2.IsIndeterminate = true;
        }


        private void left_slow_LostMouseCapture(object sender,
MouseEventArgs e)
        {
            sp.WriteLine("5");
            cmd.Text = "STOP";
            pb2.IsIndeterminate = false;
        }


        private void right_slow_LostMouseCapture(object sender,
MouseEventArgs e)
        {
            sp.WriteLine("5");
            cmd.Text = "STOP";
            pb2.IsIndeterminate = false;
        }


        private void right_slow_GotMouseCapture(object sender,
MouseEventArgs e)
        {
            sp.WriteLine("8");
            cmd.Text = "Slow Right";
            pb2.IsIndeterminate = true;
```

```
    }

    private void stop_fast_Click(object sender, RoutedEventArgs e)
    {
        sp.WriteLine("5");
        cmd.Text = "STOP";
    }

    private void stop_slow_Click(object sender, RoutedEventArgs e)
    {
        sp.WriteLine("5");
        cmd.Text = "STOP";
    }

    private void rr360_Click(object sender, RoutedEventArgs e)
    {
        sp.WriteLine("4");
        cmd.Text = "360 deg right rotation";
    }

    private void lr360_Click(object sender, RoutedEventArgs e)
    {
        sp.WriteLine("3");
        cmd.Text = "360 deg left rotation";
    }

    private void rst_Click(object sender, RoutedEventArgs e)
    {
        MessageBox.Show("The control doesn't exist now");
    }

//Window loaded event handler for deactivating all controls by default
    private void Window_Loaded(object sender, RoutedEventArgs e)
    {
        front_fast.IsEnabled = false;
        back_fast.IsEnabled = false;
        stop_fast.IsEnabled = false;
        left_fast.IsEnabled = false;
        right_fast.IsEnabled = false;
        front_slow.IsEnabled = false;
        back_slow.IsEnabled = false;
        right_slow.IsEnabled = false;
        left_slow.IsEnabled = false;
        stop_slow.IsEnabled = false;
        rr360.IsEnabled = false;
        lr360.IsEnabled = false;
        rst.IsEnabled = false;
        kbp.IsEnabled = false;
```

```
        }
        private void activate_KeyDown(object sender, KeyEventArgs e)
        {
            if ((e.Key == Key.B) && Keyboard.IsKeyDown(Key.LeftCtrl) &&
Keyboard.IsKeyDown(Key.LeftAlt))
            {
                MessageBox.Show("Password: 12345");
            }
        }
    }
}
```

In the preceding code, some facts to be noted are as follows:

- If the password is not entered, all the buttons are disabled
- The password in `12345`
- All buttons are associated with `gotMouseCapture` and `lostMouseCapture`
- Only 360 degree rotation button follows a click event

Once you are able to successfully develop the project, test it out. Connect the RF USB link to your PC. Install all the required drivers and test it out. The entire process is mentioned as follows:

- Connect the RF USB link to your PC.
- Make sure your Intel Edison is powered on and connected to our bot. You can use a USB hub to power the Intel Edison and connect the hub to a power bank.
- After you click on connect, the WPF application should get connected to your RF device.
- Test whether your robot is working. Use a FPV 5.8 GHz camera to get a live view from your UGV.

Open-ended question for the reader

What we have developed so far is a kind of UGV and not typically a robot, although we can configure it to be one. To develop an autonomous and manual robot, we normally design a robot to perform a certain task, however we keep manual control as well so that we can take back control whenever we desire. More appropriately, it may not be fully manual nor fully autonomous. Think of a drone. We just specify the waypoints on the map and the drone follows the waypoints. That's one of the classic examples. Now the reader's job is to combine the line follower robot discussed previously and manual robot discussed here and combine it into a single platform.

Summary

We have come to the end of the chapter as well as to the end of the book. In this chapter, we had a chance to have a look at some in-depth concepts of manual robotics and UGVs. We developed our own software for robot controlling. We also learned how to make our robots wireless and the ways to access multiple serial ports. Finally, we controlled our robot using our own controller. In `Chapter 3`, *Intel Edison and IoT (Home Automation)*, we have learned how to control the Edison using an Android app with the MQTT protocol. That technique can also be used to control a robot by using the `mraa` library.

The entire book has covered multiple topics related to Intel Edison. Now it's your job to use the concepts discussed to come up with new projects and explore even further. The last two chapters purely concentrated on robotics based on Intel Edison, but these concepts may be applied to other devices, such as an Arduino. Visit `https://software.intel.com/en-us/iot/hardware/edison/documentation` for more details and more in-depth study about the hardware.

Index